Farm Animal Behaviour

Cover Illustration

This illustration is from a 'stone rubbing' made by the author on a Pictish stone carving of 'The Morayshire Bull'. The carving, dated about A.D. 750, is a perfect illustration of the 'threat display' in the bull, a striking feature of masculine bovine behaviour. The carving was discovered in north-east Scotland in fairly recent archeological work.

Farm Animal Behaviour

An introduction to behaviour
in the common farm species

Andrew F. Fraser

MRCVS, MVSc (Toronto), FIBiol

Professor, Department of Veterinary Clinical Studies,
Western College of Veterinary Medicine,
University of Saskatchewan, Saskatoon, Canada
Formerly Senior Lecturer, Department of
Veterinary Surgery and Obstetrics,
Royal (Dick) School of Veterinary Medicine,
University of Edinburgh;
Director of Animal Care, University of New Brunswick,
Fredericton, Canada

Second Edition

Baillière Tindall · London

A BAILLIÈRE TINDALL book published by
Cassell Ltd,
35 Red Lion Square, London WC1R 4SG
and at Sydney, Auckland, Toronto, Johannesburg
an affiliate of
Macmillan Publishing Co. Inc.
New York

First published 1974
    Reprinted 1975
Second edition 1980

ISBN 0 7020 0776 5

Educational Low-priced Book Series Edition 1980
German edition (Eugen Ulmer, Stuttgart)
Spanish edition (Editorial Aeribia, Zaragoza)

Printed in Great Britain by Spottiswoode Ballantyne Ltd.
Colchester and London

British Library Cataloguing in Publication Data

Fraser, Andrew Ferguson
    Farm animal behaviour.–2nd ed.
    1. Domestic animals–Behavior
    I. Title
    636   SF756.7
    ISBN 0-7020-0788-9

Contents

Page

Preface to the Second Edition   vii

Introduction   1

Part I: The Nature of Behaviour

1   *Systems of the Ready State*   11

2   *Motivation, Emotionality and Drive*   21

3   *Organization of Reactions*   26

4   *Factors Affecting the Production of Behaviour*   31

5   *Acquired Behaviour*   36

6   *Environmental Influences on Behaviour*   47

Part II: The Behaviour of Development

7   *Behavioural Development*   59

8   *General Postnatal Behaviour*   66

9   *Species Development*   71

Part III: The Behaviour of Maintenance

10   *Introduction*   89

11   *Reactivity*   92

12   *Ingestion*   100

13   *Exploration*   109

14   *Kinetic Behaviour*   116

15   *Behaviours of Association*   123

16 Body Care   128
17 Territorial Behaviour   134
18 Behaviour of Rest and Sleep   143

Part IV: Species Behaviour Patterns

19 Behaviour Patterns in Horses   151
20 Behaviour Patterns in Cattle   160
21 Behaviour Patterns in Sheep   171
22 Behaviour Patterns in Pigs   179
23 Behaviour Patterns in Poultry   188

Part V: Reproductive Behaviour

24 Sexual Behaviour   197
25 Parturient Behaviour   212
26 Nursing and Maternal Behaviour   224

Part VI: Abnormalities in Behaviour

27 Stress in Animal Husbandry   233
28 Anomalous Behaviour   240
29 Behaviour and Clinical Disorders   257
Supplementary Reading   265
Glossary   275
Index   281

Preface to the Second Edition

Since the publication of the first edition of this book in 1974 much has happened in the study of farm animal behaviour. Instruction on the behaviour of the animals used in farming has been formally incorporated into the teaching programme of many agricultural, biological and veterinary institutions all over the world. In some instances this has been done in a very determined fashion. Universities have established teachers in this subject on permanent, full-time positions. Courses of substantial magnitude have been incorporated into the curricula of very many colleges to cover the subject of applied ethology, which ties together many other subjects related to animal production and veterinary medicine. In North America, for example, the subject of ethology, dealing with practical and philosophical aspects of animal behaviour in farming, is being rapidly and favourably received into the important academic world of animal science. Its future development in such custody is assured.

An international scientific journal on applied ethology has been launched, close behind the publication of the first edition, and the author has been privileged to serve as its editor-in-chief. The first world congress on ethology applied to zootechnics was held in Madrid in 1978. This was a successful rally of several hundred scientists now committed to a pursuit of domesticated animal behaviour as their primary study. Again, as the Congress President, the author was privileged to see this materialization of a scientific dream. Other major events in this subject are imminent. Developments have been so breathtaking that the author has found difficulty in going into the necessary seclusion, from the active scene, long enough to prepare this edition. It is hoped that it will do some justice to the developments described.

As before, readers are invited to use the book as a foundation for a greater study of a thoroughly rewarding enquiry into how the farm animals live their lives. It is hoped that these readers will include students, scientists, veterinarians, animal welfarists and animal farmers, together with the progressive public, who may see these animals as rather more than items of utility.

July 1979   A. F. FRASER

Acknowledgements

This textbook has received support work from numerous persons to whom the author is indebted for discussions and communications. Colleagues in many countries supplied helpful advice and valuable items of information. Dr Michael Fox of the Institute for the Study of Animal Problems, Washington, D.C. gave constructive assistance with anomalous behaviour, Dr Katherine Houpt of the Department of Veterinary Physiology, University of Cornell, N.Y. and Dr Ron Kilgour, Hamilton, New Zealand supplied useful data. Dr Henry Herchen of the Evansburg Veterinary Clinic, Evansburg, Alberta gave stimulating help in the portions of the text dealing with innate behaviour.

Most of the illustrations were produced by Mrs Anne Sheppard, formerly of the Department of Biology, University of New Brunswick and Mr Gary Cody of the Anatomy Department, Western College of Veterinary Medicine, University of Saskatchewan. Dr Rae Stricklin of this university kindly supplied the tabulated data on social dominance in cattle. Special thanks are expressed to Windfields Farms, Ontario for free access to the unique foaling facilities there, which were so helpfully provided to study equine parturition and equine neonatal behaviour.

The publishers are sincerely thanked for encouragement, gentle reminders and advice in the evolution of this book from its first edition.

A.F.F.

Introduction

Animal behaviour is the overt and composite functioning of animals individually and collectively. Behaviour is also the means whereby the animal mediates dynamically with its environment, both animate and inanimate.

Ethology is the scientific study of the behaviour of animals in their natural or usual environments. The customary environment of the domesticated animals is, of course, domestication. This study is therefore highly relevant to veterinary medicine. Veterinary ethology is the study of behaviour of domesticated species when used particularly as a means of appraising health or function. Veterinary ethology is therefore an applied science.

Behaviour in mature animals is a mixture of inherited, or innate, and acquired components. Innate components consist of simple reflexes, compound responses and complex behaviour patterns; acquired components consist of conditioned reflexes, learned responses and general habits. These various components can mix and generate a great range of behaviours.

The signs of behaviour are often specific signals or 'releasers'. Collectively they have been popularly termed 'body language'. Behavioural signs induce responses in other animals. By a system of signs and countersigns many animals interrelate, their behaviour being determined by physiological and neurological needs and triggered by information coming to them from their environment, both animate and inanimate.

In 1973 three scientists received the first Nobel prizes in ethology. These pioneers of ethology were Lorenz, Von Frisch and Tinbergen. Formerly two schools of thought divided the science of animal behaviour. Europeans, calling themselves 'ethologists', concentrated on instinctive behaviour, observing animals in the wild. The American school of 'psychologists' was more interested in behaviour under controlled laboratory conditions. Both recognized Konrad Z. Lorenz as 'the father of modern ethology'. Lorenz laid the groundwork for many of the most important lines of research still pursued today. He formulated a new school of investigation, based on his conviction that an animal's behaviour, like its physical adaptations, was part of its equipment for survival. He proved his point with a wide variety of animals. Among the many basic truths shown by his experiments, all conducted with animals roaming freely in their natural surroundings, were such learning processes as the phenomenon of imprinting. In this, the newborn animal has its first environmental conception firmly impressed upon it to motivate and direct its behaviour subsequently. Such discoveries in animal behaviour and their approval have helped ethology to become recognized as an established science.

All sciences have their disciplinary constraints. Ethology has its full share of scientific standards to meet. These cannot be compromised, even under the pressure of responsibilities to goals. More research work must obviously be done. It begins to appear likely that animal production scientists will engage increasingly in ethological pursuits, such as animal sociology. Productionists are certainly showing fresh and serious interest in the whole subject of food-animal behaviour. As a group, they represent huge potential resources of appropriately trained and experienced scientists. Minor shifts in philosophies, technical interests and research commitments would be the only changes required, on the part of many scientists, for ethological participation with veterinarians on this common ground.

Ethology recognizes motivated behaviour in animals. Such behaviour implies the existence of animal emotions. Many of these emotions and feelings are recognizable. These include pain, hunger, fear and rage. Other emotions are also evident in animals, although it cannot be assumed that these exist in forms closely resembling those experienced in humans. Excitement is one. Excitement depends upon perception and senses. Many animal ethologists readily concede that animals have anatomical and physiological equipment similar to man to facilitate the experience of sensations. Some animal senses are now shown to be much keener than the equivalent ones in man. Between people, the range of sensory perception and associated emotion is great. Among animals and species, it is certain that the range will be no less. Denial of feelings and emotions in animals would be scientifically dishonest today. Acknowledgement of emotions, albeit different in degree and quality from those of man, is reasonable and logical. Recognition of the communication of sensations between animals is one of the principal achievements of ethology. The pioneer ethologists recognized 'signal codes' in the behaviour of animals and an early common philosophy among ethologists was that, where the 'signal codes' of a given species of social animal could be recognized as a language, the behaviour could be understood by an observer who had learned the behavioural 'vocabulary' of the species. While there is no speech among animals, in the human sense, an extensive repertoire of meaningful vocalizations is becoming recognized.

Ponderous investigations into the concepts of pain and stress in animals have taken place. Some of these investigations have confirmed that a problem of human understanding exists. Others have shed light on this problem. The contemporary concepts of pain and stress in animals cover three states which are interpreted simply and briefly as follows. (1) The condition of discomfort relates to any adverse interference with the animal's normal state of health or well-being. (2) Stress is a physical condition with extensive manifestations revealing undue 'tension' or 'anxiety' relating to environmental factors. (3) Pain is recognizable by more positive behavioural signs such as struggling, screaming, squealing and convulsions. The fundamental ethics of veterinary or applied ethology would not permit the calculated and unnecessary induction of such states.

Significant behavioural needs can be recognized. Some of these exist in behaviours of maintenance. Kinetic behaviour is an innate behavioural need, not always met under intensive husbandry conditions. The behaviour of 'association' shows the extent of social activities in the total system of a species' behaviour. Many of these needs cannot be met in confinement. In animal sociology, behaviour in categories such as communication, aggression, territorialism, dominance and symbiotic relationships has been comprehensively reported, showing the rich fabric of behaviour in social animals. So much highly organized social behaviour is now recognized, in farm animal socio-ethograms, that this can be seen as a major part of the genetic programming of their species. Richly patterned innate social behaviour represents a major behavioural output. Social opportunity can therefore be recognized as an ethological need in the interests of the organism's functional integrity.

Ethology embraces various sciences. Its multidisciplinary character is emphasized in applied studies. Among ethologists as a body, the effective blend of various disciplines demands some basic agreements, some common philosophies, some intellectual harmony. In addition ethology, again most notably in its applied form, is also a vocation. This latter fact puts certain other demands on its practitioners individually. Those who work in applied ethology are generally motivated by a desire to participate in the utilization of knowledge in the interests of life. Students of applied animal ethology are not unaware of their vocational connection with the subject. They also appreciate fully that this can best be served by utilizing the principles of science. Ethological communication has now become well developed through its own press and symposia. This communication has moulded the philosophy and developed the ethics of applied ethology.

Although it is truthfully contended that many of the veterinary subjects taught already embrace aspects of animal behaviour, it is clear that ethology must be introduced formally and compulsorily in the preclinical veterinary curriculum. Some schools have already led the way by providing short series of lectures, introducing the principles of the study of animal behaviour. It is now found that an increasing proportion of new students come from urban backgrounds and do not possess a practical knowledge of animal behaviour as did the majority of their predecessors, and for this further reason there is need to include instruction in animal behaviour in the courses leading to qualification of the modern veterinarian.

Practical ethologists intend their work to facilitate progress in several directions. They know that ethology will rationalize animal care. They know that ethology will promote optimum utilization of animals collectively and individually. They know that other subjects, such as ecology, animal production, husbandry, reproduction, veterinary science and branches of biology, will benefit from ethological process.

THE ROLE OF ETHOLOGY

The study of animal physiological systems and the functions of the discreet anatomical items is undoubtedly of enormous help to the student of ethology. But those studying the workings of animal parts are the first to admit that this knowledge, however detailed, does not provide a satisfactory understanding of the mechanism of the whole. Conversely, there are other biologists who, having studied animal behaviour, may speculate on the function of certain parts, such as the brain. This approach is no more satisfactory than the first. Clearly, what is needed is a broad, basic understanding of physiology and anatomy together with numerous supporting disciplines, such as mathematics. Not least of all, a general knowledge of animal recognition is required if an ethologist is to be competent.

The acquisition of knowledge alone has justified scientific studies on a massive scale. Ethology, as a scientific discipline, can justifiably be studied for its own sake. Almost certainly the earlier students of the subject, and many current ones, had no other motive in selecting this subject as a major interest. Gradually, however, more pressures are being put on ethology students, both by animal husbandry scientists who wish to modify the ways in which animals are husbanded and by that section of the informed public which has acquired some conscience about the manner in which animals are utilized by mankind.

Nowadays animal scientists must all have a working knowledge of animal behaviour. It is impossible, though, for any one person to acquire a satisfactory depth of knowledge about the behaviour of several species of animals. One therefore sees increasing numbers of students being forced to concentrate on a limited number of species. The species of main concern now are farm animals, namely horses, cattle, pigs, sheep, goats and poultry. The public concern over animal welfare is similarly directed largely at these species. The farm animal scientist is now expected to be able to provide scientific information about the behaviour of these species, as a result of sound instruction and personal observation. Modern farm animal scientists, whether they like it or not, are expected by the majority of people therefore to be competent to act as experts on matters concerning farm animal welfare. Indeed as the farm animal industries become more technologically minded, more stock-keepers are prepared to accept informed comment on animal behaviour and its relevance to animal welfare.

ETHOLOGY AND ANIMAL WELFARE

The Brambell Report placed emphasis on the lack of behavioural studies on the welfare of animals kept under intensive methods of husbandry. It stressed that these studies were important not only for the animals concerned but also for the welfare of the animal industry. The Report questioned the soundness of taking one single parameter of production, such as growth rate, egg or meat production, as the only reliable measure of the suitability of a system of husbandry. It pointed out that any sufficient estimate of an animal's welfare must not be based only on the physical evidence of productivity. The Brambell Report defined as unworthy those arguments, however plausible, which implied that animal husbandry practices in operation cannot be interfered with on welfare grounds without scientific proof of suffering. The committee insisted that animals show unmistakable signs of pain, exhaustion, fright, frustration, rage and other emotions, any one of which may indicate a degree of suffering. Further, it pointed out that experienced stockmen were able to detect unusual signs in their animals and to appreciate the implications of these with regard to suffering. Modern stockmen are no less well-meaning but modern intensive methods of animal productivity demand a greater control over the animals and their environment. This control includes limitations on space, diet, ventilation, illumination, bedding and companionship. As the Brambell Report pointed out:

> It is because of these various restraints imposed on the animal by the artificial environment of intensive husbandry that the role of welfare has become so crucial today.

The committee's report was summed up by the following statement:

> Above and beyond all these matters stands the fact that modern intensive animal production methods most markedly increase the responsibility of those who use them towards the animals in their charge. If any creature is wholly and continuously under control, we believe that this total human responsibility must be acknowledged; changing patterns of husbandry may mean varying degrees of frustration and discomfort to animals whose normal patterns of behaviour are still imperfectly understood. We are certain that a beginning must be made to safeguard their welfare.

Animal scientists have come to realize that there are many ways in which ethology can help with the welfare of animals and thereby ensure more profit from more sensible husbandry. For example, comparative studies on feral animals help to throw light on the behaviour patterns of domesticated species. Studies of this kind have been been helpful in clarifying the importance, in the lives of modern farm animals, of such environmental factors as social companionship, regular feeding and resting regimens and the rhythms of other activities such as sleeping and breeding. For ethology to continue to contribute substantially to profitable animal husbandry, more and more exponents are required to fill the gaps in knowledge. Where can we better apply animal ethology than in those species which are totally under our control and totally dependent on us for sustenance and existence?

The breaking-up of behaviour patterns into component units, which can logically be considered in isolation, is necessary in attempts to specify behaviour. The obvious interrelationships between these units encourage the postulation of the various concepts, which together become the theory of the subject. Such theory generates its own technical glossary and the extensive acceptance and use of this common vocabulary improves the exchange of technical ideas among students of the subject. The fact remains though that, however valid various hypotheses might be, it is important to realize that theoretical concepts are principally of use in the teaching of the subject. They are, therefore, a means to an end: the end being a comprehensive appreciation of the significance of behaviour.

Part I
The Nature of Behaviour

I *Systems of the Ready State*

An animal's behaviour is the result of the complexity of interactions which take place between it and its environment. It is to be seen as the overt and composite functioning of the whole animal—a definition which has already been offered in this book. Early in the study of behaviour then, it is necessary to look at some of the mechanisms by which animals function, namely, their physiological processes. In reviewing the principal physiological features of ethological consequence, foremost attention must be paid to the role of sense organs and the neural and endocrine mechanisms.

NERVOUS SYSTEM

Information about internal and external factors concerning an animal passes to the central nervous system through a range of receptors. Stimulation affects the animal's sensory receptors and may be varied in nature. It can be chemical, thermal or mechanical. The sense organs convey a received stimulus to the peripheral nerves which in their turn produce the phenomenon of excitation. When stimuli have been built up to an adequate level they provide a train of nerve impulses in the form of waves of electrical activity sweeping along the surface of the individual nerve fibres. Nerve impulses are of the same charge at all times but differ in their frequency. When the stimuli increase in strength the frequency of the impulses increases, and although responses are not constantly related to the impulse they nevertheless tend to be.

The receptors which provide information for the nervous system are of two main types: exteroceptors and interoceptors. Exteroceptors deal with stimulation originating outside the animal. They include the receptors found in the skin as well as the organs of vision, hearing and smell. The latter organs are sometimes referred to as teleceptors since they can deal with stimuli originating some considerable distance from the animal. Interoceptors are concerned with stimulation originating within the body and are located within muscles, joints, tendons and internal organs.

Receptors which provide nervous impulses giving rise to sensations in the animal are known as sense organs. The principal sense organs in determining animal behaviour include the organs of sight, smell and hearing. Other receptors create impulses which do not reach consciousness but nevertheless produce reflexes. Such receptors are called activators.

Of fundamental importance to the transmission of impulses through the nervous system is the synapse. When an impulse passing along a nerve fibre reaches the fibre's end and meets another neuron (or nerve cell), it crosses over at a synapse. Synapses are the points of contact between neurons. Trains of impulses arriving at a synapse are passed on if they are of sufficient frequency. The function of the synapse is to control neural transmissions, limiting cross-communications between nervous tracts. Some synapses have an inhibitory effect on nerve signals. Synapses connect neurons to form specific neural pathways. In many cases the neural paths lead to the cerebral cortex which is the outer covering of the cerebrum.

Nerve impulses are also received into the brain via the reticular activating system which transmits signals concerned with fundamental physiological affairs. The dorsal columns of spinal nerves convey impulses received from touch receptors on the surface of the animal's body.

Cerebral Cortex

The cerebral cortex, although acting as a unit, has certain localized regions where sensory impulses are received and subjected to redirection. These specialized areas of the cortex are primary sites for sensory reception, the nervous activity subsequently spreading over a greater area. The cortex has a multitude of cells and neural paths, each one communicating with many others. This extremely complex relay system permits tremendous variability in the way that nerve impulses may be channelled.

The cerebral cortex possesses four main sensory areas into which projectory fibres discharge. These are:

1. The somasthetic or body sense area
2. The visual area
3. The auditory area
4. The olfactory area

All of these are important in the receipt and interpretation of nerve signals, and are fundamental in determining behaviour.

Somasthetic area. This area is sited in the parietal lobe of the cortex and receives nerve impulses from very many part of the body including its surface.

Visual area. Nerve fibres are collected from the retina of the eye into the optic nerves and are distributed, within the cortex, to the extensive visual area at the occipital part of the cerebrum. Recognition of patterns and releasers takes place in the visual area.

Auditory area. This is located in the temporal lobe of the cerebral cortex. The area receives nerve impulses concerned with auditory sensations from the thalamus. The fibres of the hearing nerve end in the pons, from which region other fibres pass to the thalamus and then to the cortex.

Olfactory area. The sensory area dealing with smell plays a much more important role in breeding behaviour than has generally been recognized. An olfactory region is located in the hippocampus which receives projection fibres from the centre in the olfactory bulb. This centre deals with olfactory reflexes. The fibres of the olfactory system originate with nerve cells located in the mucous membrane of the nasal passages and terminate within the olfactory bulbs.

It seems that areas of the cortex are designed to correlate dynamically with receptive areas on the body's surface. The neural importance of the superficial area and central nervous areas changes according to the major activities of an animal; for example, it seems that the appropriate area of the cortex is more receptive to stimulation from the genitalia during breeding times. Some regions of the cortex communicate directly with the hypothalamus; these are the 'old brain' regions, e.g. the areas of the frontal lobe and of the orbital frontal regions. Various part of the forebrain convey neural activities to the hypothalamus through a variety of systems, either directly or after interruptions, so that by one route or another it is in receipt of impulses from optic, olfactory, acoustic, tactile and internal sources.

The limbic system, located deep in the subcortex, apparently contains neural centres such as the amygdala which control aggressive behaviour in its various forms.

Hypothalamus

It is at the level of the hypothalamus that patterns of nervous activities become integrated and regulated so as to establish the adaptive reactions of the animal. Even behaviour which is largely dependent upon experience and learning in the animal is seldom, if ever, completely free of control by the primitive mechanisms established in the hypothalamus and in the subcortex generally. The neural links involving the subcortex, and the links between the hypothalamus and the surrounding brain in particular, remain the principal integrators of most behavioural patterns.

The working units of the hypothalamus are neurons which are grouped into 'nuclei'. These nuclei operate together in a fashion resembling a computer. The information from various levels of the brain is received and processed by these nuclei before signals are subsequently reissued to more specialized parts of the body, which are geared to function under the control of the hypothalamus. Much of the influence of hypothalamic activity is directed at the production of hormones in the subjacent pituitary gland. The pituitary is the principal endocrine gland in the body and its hormonal production is all-important in the maintenance of the bulk of the body's activities, including behaviour. Even the all-important central hypothalamus is responsive to some of the endocrine activity for which it is initially responsible. Quite recently it has become clear that there is probably a considerable amount of hormonal control over the hypothalamus. The receipt of afferent stimulation gives the hypothalamus the role of maintaining and regulating the activity of the pituitary gland.

ENDOCRINE SYSTEM

The nervous system and the endocrine system are clearly adapted for different roles but contact between them is essential for their full function since they are interdependent. The two systems cooperate with each other through the processes of neural secretion and through the effects of hormones on the brain.

Hormone secretion is subject to the influence of many forms of stimulation. Functional endocrinology now recognizes an elaborate organization of interactions between the animal's own activity, the external stimuli which it receives and its internal physiological state. Any of these three factors—behaviour, environment and internal state—can alter in force to cause a change in the others. This elaborate apparatus clearly creates a potentially complex situation.

To obtain some understanding of this complex mechanism, the function of the individual endocrine glands requires examination.

Pituitary Gland

The pituitary gland is of first importance since it exercises a primary role through its control of other endocrine glands which are, in their turn, secondary within the endocrine system.

The pituitary gland—consisting of two separate regions, the anterior and the posterior lobes—is suspended from a stalk at the base of the brain, at the rear of the cranium and is closely connected to the hypothalamic region. A nerve tract originating in the hypothalamus runs through the pituitary stalk to feed the posterior lobe, while the anterior lobe of the pituitary gland communicates with the hypothalamus through a portal system of blood vessels. This establishes a direct link between the hypothalamus and its cells and the pituitary gland.

The two portions of the pituitary gland produce and secrete specific hormones which act as chemical messengers circulating to other endocrine glands in particular and to distant parts of the body, known in this context as target organs, which may in their turn produce their own hormones.

Of great importance in reproductive behaviour is the output of the gonadotropic hormones from the anterior pituitary. These hormones are the *luteinizing hormone* (LH) and the *follicle-stimulating hormone* (FSH) which are produced in the male and in the female in concert and influence the activities in the respective gonads, the testicle or the ovary.

In the female, the two hormones, although produced together, dominate each other alternately; this gives rise to the cyclic activity of reproduction characteristic of the female. In the male animal, however, the production of these two hormones appears to be level and continuous.

The posterior pituitary gland produces *oxytocin*. This important hormone is responsible for various uterine activities and also the release of milk in the lactating animal. It is this hormone which encourages the outflow of milk at suckling. Oxytocin is produced in the maternal subject in response to nerve signals received following tactile stimulatory behaviour in her mammary region on the part of nursling animals. Visual and other stimuli are also provided by the nursling's appearance and behaviour. Sexual stimulation of the male is also known to be attributable to oxytocin production.

Thyroid Gland

One of the glands influenced by the pituitary is the thyroid gland and the hormone which it produces is *thyroxine*. This is closely involved in most of the body's activities concerning energy output and therefore affects behaviour in general. The steady production of thyroxine is important in maintaining regular maternal activities, female cyclic sexual activities and sex drive in both sexes.

Adrenal Gland

The adrenal glands—one adherent to each kidney—produce steroid substances in their outer cortices. Production is a consequence of hormonal stimulation from the pituitary gland. In the male, these steroid substances are termed *androgens*. Androgens are the male hormones and those from the adrenal act in support of the principal male hormone produced in the testis. The effect of adrenal hormones on sexual activity must be considerable, however, since it is known that disease of the adrenal cortex can cause marked alterations in sexual behaviour through altered production of adrenal hormones.

The central portion of the adrenal gland, namely the medulla, is the region in which *adrenaline* is produced. The sudden release of adrenaline into the circulation is associated with marked sudden changes in behaviour. Threat behaviour, flight behaviour and behaviour combining these, such as the fight-or-flight reaction, all result from a sudden increase in adrenaline production. Again a sudden increase in adrenaline output results in agonistic behaviour in general and is likely to promote forms of behaviour such as fighting among male animals and maternal protective behaviour. Adrenaline provides the basis of all the alarm reactions and it is also known that increased adrenaline output arrests the flow of milk in the suckling animal. Some release of adrenaline from the medulla of the adrenal gland proceeds almost continuously, but sudden increases in its secretion are known to result from distributing stimuli in the immediate environment of the subject.

Sex Organs

Testis. The testis of the mature male secretes *testosterone*, the principal hormone responsible for most of the typically male behaviour. In those species where there are seasonal fluctuations in the expression of male sex drive, such as rams, there is a corresponding cycle in the production of testosterone.

Ovary. In the female animal the ovary is the source of *estrogen*, the hormone responsible for female sexual behaviour. The ripened follicle of the ovary which precedes ovulation is the principal source of this ovarian hormone. The follicular cells increase in size rapidly just before estrus. At this time the output of estrogen increases greatly. Once the estrogen in the animal's circulation has reached a threshold level, all the behavioural signs of estrus typical of that species are shown. Estrous behaviour is maintained for the period during which the follicle is at the peak of its maturity, but when ovulation occurs with the bursting of the follicle a sharp drop in oestrogen production occurs and estrous behaviour ceases.

Hormone levels. With both estrogen and testosterone, there is no clear relationship between the quantity of hormone and the intensity of the sexual behaviour shown. An adequate level of estrogen in the body will only ensure that estrous behaviour will be exhibited. The degree to which this behaviour will be shown is not dependent on the hormonal level, but is principally under the control of inherited nervous factors. In both sexes a feedback mechanism operates between the hypothalamus and the gonad. The level of sex hormones produced by the gonad is communicated to the hypothalamus. The hypothalamus is capable of reducing the endocrine activity of the gonad if the level of sex hormone becomes excessive; likewise, if the level of sex hormone is insufficient, the hypothalamus can quickly stimulate the pituitary into increased gonadotropic output.

Pineal Gland

The pineal is a very small endocrine gland located deep in the brain and directly above the hypothalamus. In some lower forms of animal life this gland is known as the 'third eye'. In the large farm animals the pineal gland produces a hormone called *melatonin*. Melatonin appears to control cyclic breeding behaviour. It is evidently under the control of light stimuli and is, therefore, a form of third eye even in mammals. Certainly estrous behaviour in animals requires melatonin production and melatonin in its turn requires the stimulation from light and dark phases.

Hormones in General

It has been well known that hormones exercise a great deal of control over behaviour. Today, the improved knowledge of endocrinology justifies this general belief by showing the many ways in which hormone production takes place in animals. In many cases this production is under external influences and ultimately dictates that behaviour which is appropriate for an animal in its own environment.

Endorphins are brain hormones which have opiate properties, being able to kill severe pain in distant areas of the body. In addition, it would appear that such 'brain hormones' can also control emotionality, increase concentration and improve behavioural attention to problematic situations.

STIMULATION

Some tactile sensations are clearly pleasant to animals. Grooming activities, for example, are largely concerned with the exchange of tactile sensation between pairs of animals. Individual grooming activities also depend upon the reverberation of the tactile sensory apparatus.

Olfaction. The sense of olfaction is of critical importance in the stimulation of a wide variety of responses in animals. Reproductive responses, for example, are quite evidently under the control of the olfactory senses to a very large extent. Odour can be seen to have a stimulatory value in arousing the male sex drive. Odorous substances eliminated by one animal which have the specific effect of stimulating another animal, usually of the opposite sex, are termed *pheromones*. The importance of the role of pheromones in the breeding behaviour of animals is becoming much more widely acknowledged. These pheromones have various sites of production and routes of elimination. For example, they are produced in the preputial fluids of the boar. There are a variety of male animal odours which are detectable even to man; the smell of the billy goat, for instance, is almost certainly pheromonal in its effect on female members of the same species.

The production and the reception of odour are clearly important in generating behaviour. Odour plays a large part in the establishment of the strong bonds between a mother and the newborn animal. These bonds are dependent firstly on mutual recognition through odour.

Visual stimuli. Although odour is the principal means by which early recognition occurs between mother and young, visual recognition soon takes over as the secondary means of mutual identification.

It has been known for some time that the relative length of the light period of each day is a factor in determining breeding behaviour in farm animals. Seasonal breeding, for instance, is largely determined by the changes in the daily photoperiod. Photoperiodism operates in two principal ways:

1. Some animals exhibit their reproductive activities during that portion of the year during which the daily light period is long. It is widely known that for horses the normal breeding season commences in the spring—that period of the year when light is becoming stronger and the number of daylight hours greater—and continues through summer.

2. Some animal species confine their breeding behaviour to that portion of the year characterized by the minimum amount of daily light; sheep and goats are examples of this. Most breeds of sheep and goats commence their breeding seasons in the autumn, when the daily photoperiod is less than the dark period and the light period is diminishing further day after day.

Clearly, the natural light stimulus for those farm animals that show seasonal breeding is a complex one involving the absolute quantities of light and dark as well as relative quantities of light each day which are changing dynamically. Although it is generally believed that daily fluctuations in the photoperiod emphasize the change taking place in daily light rations, it is also clear that the fixed nature of the photo-period is important, i.e. seasonal breeding animals maintain their breeding activities as long as an adequate quantity of light (or of dark) is delivered. When the photoperiod fails to provide adequate stimulation for the animal a refractory period develops during which the breeding performance is arrested.

Auditory stimuli. Vocal expression by animals ensures that auditory stimuli are being exchanged. Auditory stimulation has been studied in horses and pigs. In the latter species most study has probably been devoted to the effect of auditory stimulation on breeding. There is abundant evidence that auditory stimulation plays a large part in maintaining the close bond between dam and the newborn animal.

Clearly animals react to auditory stimulation of a great variety of types. They show alarm in response to certain sounds while other sounds produce an evidently reassuring effect.

Gustatory stimuli. Taste plays an important part in grazing behaviour and ensures that the requisite elements are consumed by an animal in appropriate quantities to maintain health. Animals suffering from various forms of nutritional deficiencies usually first give evidence of this by abnormal feeding activities which indicate that palatability has been altered for them.

2 *Motivation, Emotionality and Drive*

An animal's motivation can often be deduced from its behaviour. When behaviour leads to a recognizable goal the motivation is identifiable. Motivation stems from the blending together of numerous items, such as stimulating factors, the current physiological status and the animal's characteristics from inheritance and experience. These various elements obviously include innate ones, but all may blend variably and motivational change results. Motivation typically exhibits specific direction. In ethology certain specific motivations are termed drives. Because of the clarity of motivation, through the detection of goal or intent, the sexual, maternal and feeding drives are those most readily appreciated. Other drives can be recognized, however, including the drives of play, exploration and the main systems of self-maintenance behaviour. The concept of drive is useful in ethology.

MOTIVATION

Motivation is the process responsible for the goal-directed quality of behaviour. Much of this behaviour is clearly related to maintenance. In such goal-directed behaviour, specific bodily needs are satisfied. Needs have a correlate in need satisfaction. The neurophysiological integration of much homeostatic goal-directed behaviour is discussed with regard to the behaviours of ingestion, body care and reproduction. Many kinds of motivated behaviour have little apparent relation to physiological homeostasis, but it is a useful generalization that motivated behaviour is induced by needs and is sustained until the needs are satisfied. Inseparable from motivation is the concept of reward and punishment. Rewards are things that animals work for, or things which strengthen behaviour, and punishments are the opposite. These are related to motivation in that rewards may be said to lead to satisfaction of need. Although some rewards and punishments have conscious correlates, many do not. Accordingly much animal behaviour is influenced by factors of which the subject is presumably unaware.

The brain area most concerned with the integration of motivation behaviour related to homeostasis is the hypothalamus. It contains the integrating centres for ingestion, body care, etc. Areas of the hypothalamus normally mediate highly motivated behaviour such as ingestion and reproductive activities. Activation of the neural pathways underlying these behaviours is in itself reinforcing and can provide motivation to engage in them.

EMOTIONALITY

The complex phenomena collectively termed emotionality are related to motivation. Much experimental evidence points to the involvement of the limbic system in displays of emotional behaviour. The limbic system is an interconnected group of brain structures within the cerebrum. It includes portions of the frontal lobe cortex, temporal lobe, thalamus and hypothalamus, together with the linking neuron pathways connecting all these parts. These parts of the limbic system therefore have many connections with each other and with other parts of the central nervous system (CNS). It is likely that information from all the different afferent routes influences the limbic system. Activity of the limbic system can result in a wide variety of autonomic responses and body movements which comprise behaviour.

Three distinct neural systems mediate the various emotional behaviours of the limbic system. Following experimental stimulation of one area the animal actively approaches a situation in a positive, exploratory manner. Stimulation of a second area causes the animal to stop any behaviour it is performing. Stimulation of a third physiological area of the limbic system causes the animal to show marked aggressive behaviour typical of an enraged or threatened animal. An animal's behaviour can be changed, experimentally, from a quiet state to savageness, or from the latter to docility, simply by electrical stimulation of different areas of the limbic system. Destruction of a nucleus in the tip of the temporal lobe produces docility in an otherwise savage animal. A rage response can also be caused by destruction of part of the hypothalamus. Stimulation of certain hypothalamic areas of limbic structures elicits behaviour in animals which seems to have a strong 'emotional' component. The medial hypothalamus exerts inhibition on the circuits producing fight-or-flight behaviour. Upon receipt of appropriate environmental stimuli, the temporal lobe inhibits the medial hypothalamus, allowing activity in the system to increase. The resulting emotional behaviour is then the result of limbic integration. Although the structures involved in the control of emotional behaviour are predominantly located throughout the limbic system, the main controlling centres for consummatory or drive behaviour are located in the hypothalamus.

DRIVE

The classically simple account of physiological drives is that they are 'maintaining stimuli'. They persist, or are maintained, to keep the organism appropriately active until a goal is reached. The primary drives relate, evidently, to the behaviours of maintenance. It seems that drives are powerful energizing forces created by 'maintaining stimuli' which stem, in their turn, from physiological needs and activate animal behaviour towards appropriate goals. Goal attainment creates drive suppression and need reduction.

Recent research in neuroscience has revealed the substantial potential for modification of drives centred in the hypothalamus and the role of the cortex in remodelling drives has become apparent. Positive drives appear to use dopamine neurons and slow-transmitter catecholamine-permeated pathways. With the loss of such pathways, from lesions, the cortex can support drive behaviour. Apparently a repertoire of drive behaviours can be established in the cortex after the hypothalamic originators of these have been removed. A striking population of neurons in the catecholamine path through the hypothalamus consists of large neurons with outspread axons and these seem to be drive neurons. The broadcast fibres of these peptide transporting neurons, after passing through the hypothalamus, form networks of terminals sufficiently diffuse to explain the relative lack of localization or restriction of the sites effective for drive support. Thus it might be that proteins carried to the cortex by catecholamine neurons would affect drive and reward behaviour when taken up by the cortex storage elements. Improvement in drive behaviour has been produced by lesions encouraging such developments. Drive mechanisms of the brain might be mainly peptide events, that is changes in the chemical state of the brain due to hormones. Peptides, which might carry drive messages, could be carried through the brain from hypothalamic stations by catecholamine fibres and on release could produce drive states. A drive message could be carried by a pattern of activity that would release peptides. The catecholamines seem to help trigger hormone events and then the hormone process seems to carry the momentum over longer periods.

The sex drives in both sexes are manifest in a vast array of behavioural features. Essentially the sex drive depends on a given level of hormone being produced and acting on the neural tissues in the presence of appropriate environmental stimulation. Specific centres in the central nervous system have been recognized as being totally concerned with sexual behaviour, but the mechanisms are imperfectly understood in farm animals. The production of the sex hormones by the gonads is, however, satisfactorily understood. The effects of these hormones, in their very small quantities, are impressive.

The neurochemical basis of the maternal drive is poorly and inadequately understood and remains speculative and conceptual. The drive appears to be partly the result of an appropriate mixture of reproductive hormones entering the circulation at and subsequent to parturition.

The hunger or feeding drive has a physiological basis which is well documented. Hunger is evidently derived, principally, from the actions of opposing hunger and satiety centres located in the hypothalamic region. These centres operate on the basis of neural information from the gut and from glucose and other biochemical levels in circulating blood. These centres control food intake and damage to them causes abnormally excessive or deficient appetite.

Changes in levels of drive are to be seen under certain typical circumstances of season and husbandry. Variations in drive among individuals can be observed. This draws attention to innate factors in behaviour and the role of genes and DNA in animal behaviour. In a free state farm animals aggregate in groups of their own kind. Through systems of social organization, group harmony is a prominent feature of collective behaviour. Phenomenal features of group behaviour are synchrony and conformity of drive-related activities. Group drives are evidently of a higher order than drives of individuals, so that the discipline of synchronous conformity is imposed on the maintenance behaviours of a group. The discipline shows in the way that a herd, or a flock, will feed together, move together, react together, associate together, utilize territory together, shelter and rest together. Group behaviour shows cohesive drives which often appear to be states of higher motivation than would normally be observed in the behaviour of the solitary individual. Examples are seen in feeding, kinesis, body care, territorialism and resting. Reproductive behaviours, too, appear to be influenced by group or population dictates.

3 *Organization of Reactions*

Various forms of behaviour are seen to be the result of neural function within specific anatomical regions of the CNS which represent levels of control in the hierarchial system just described. Examples are as follows:

Cord behaviour. Simple reflexes such as kicking in response to a local limited stimulus have cord organization.

Diencephalic behaviour. The subcortex yields aggression, typically induced by very intensive, highly comprehensive stimulation of an aversive nature. An example would be active removal of a young nursing calf from its mother. After the removal, the cow would be likely to show a variable repertoire of aggressive behaviour. The diencephalon is the major reservoir of behaviour such as flight, feeding, sexual responses and agonistic activities.

Cerebellar behaviour. This is seen as fine adjustment of motor activity issuing from other motor centres in the CNS. The cerebellum is not itself a reservoir of programmed behaviour patterns, being limited and highly specialized in its role of a fine adjuster of motor activity. Loss of cerebellar control leads to ataxic behaviour.

Cortical behaviour. This manifests itself in acts of voluntary quality and comprehensive quantity. Representative of this would be much of the organized social behaviour in hierarchial systems such as a functional and 'harmonious' peck order in animal groups. Leadership and followership in a herd of cattle provide an example of this.

EVOLUTION OF HIERARCHY IN THE CNS

Parallel evolutionary hierarchy of the CNS is manifested by the rising complexity of behaviour patterns from the lower to the higher vertebrates. In poultry, for example, there is no evidence that there is anything other than cord control of excretion. In birds, eliminative acts are strictly spinally controlled and are not embodied in multi-purpose eliminative behaviour patterns. The behaviour of birds in general is more rigid and less variable than that of mammals. This is due to the paucity of advanced development of superior neurological control. In general, we may say, the more neurons within the neurological hierarchy become involved the more flexible are the patterns of behaviour. The higher the evolutionary standing, the more neurons become involved in behaviour.

Such flexibility of behavioural responses is advantageous to adaptation in the face of environmental variability. Low reflex spinal arcs are by no means eliminated or replaced but, in domesticated mammals, for example, they are modified and controlled by superior reflex arcs involving the cerebrum.

A high order of control of defaecation is seen in the eliminative behaviour in horses generally and breeding stallions in particular. The patterns of defaecating behaviour in horses are often elaborate and specialized. At the same time, such patterns are not always operant: for example, if the horse is being worked. The demands of this work engage the cerebrum sufficiently to return the control of defaecation to the appropriate spinal centres. This example again demonstrates that lower levels of CNS control are never eliminated or replaced by higher ones. They are only made subordinate in the hierarchial order as it has evolved.

CNS INPUT AND OUTPUT

Sensory input into the CNS will result in behavioural output, generally speaking. The exchange of stimulation and response, behaviourally, is determined by the polarity of the neuron as the basic building block of the nervous system. The peripheral receptors of stimulation are many and varied and include highly specialized sensory organs providing environmental information. Response to this information, in behavioural terms, is equally variable, ranging from motor responses through overt secretory activities to social adjustments.

If normal patterns of behaviour are suppressed, an alternative mode of expression will develop. This rule is exemplified by the way 'ethostasis' results in anomalous behaviour. For example, race-horses when confined will often adopt a stable vice.

If the environmental input is reduced, innate motivation will remain sufficient to generate behaviour, usually of a stereotyped form. Battery-confined chickens, for example, commonly indulge in persistent feather-picking.

NEUROLOGICAL PRODUCTION OF BEHAVIOUR

The neuron is the constructional unit of behaviour and contains the design of behavioural patterns. Behavioural information therefore resides in the neuron. Nissl substance, abundantly present in the cytoplasmic reticulum of the neuron, may be suspected to be the storage of 'behavioural potential'. This substance, composed of large granular bodies of ribonucleoprotein, is seen to be reduced in amount after animals have been behaviourally exhausted. There may indeed be a relationship between the ways in which genetic information and behavioural potential are stored. The quantity of RNA in the neuron

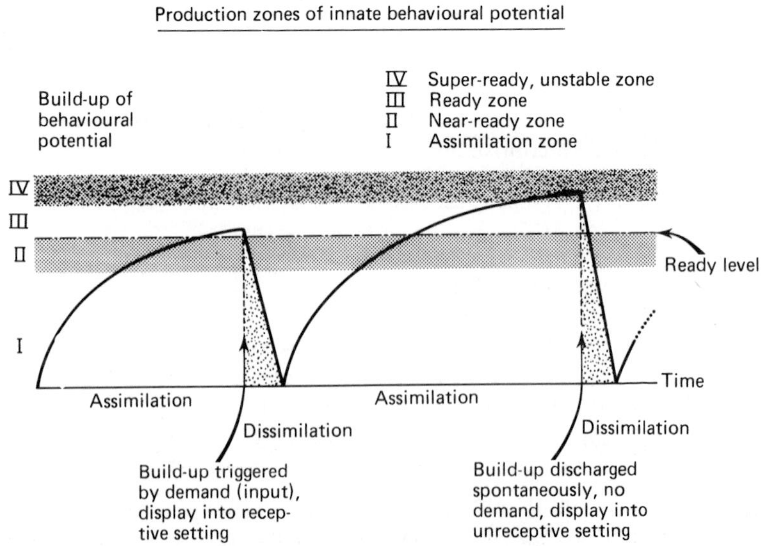

Production zones of innate behavioural potential

Fig. 1. The synthesis of behaviour.

parallels the degree of behavioural potential. No other tissue contains such heavy concentrations of RNA as does the neuron, by means of its Nissl bodies. The function of the Nissl bodies would seem crucial to the production of behavioural patterns. The RNA dependence on DNA, so well demonstrated in the field of genetics, is suggestive of a further linkage between Nissl bodies and nuclear DNA of the neuron. The aggregation of neurons constituting the nervous tissue and neural system has a massive potential for behavioural production corresponding to information input. The sensory information input is dealt with by the neural cytoplasm, rich in RNA protein.

Recent research in neurophysiology, in neurohistology and, in particular, on the ultrastructure of the neuron indicates that the Nissl bodies, which occupy much of the neuron body's protoplasm, consist of masses of ribosomes packed along the edges of the elongated lacunae set out in groups of parallel rows. These evidently communicate with the tubule running the length of the neuron's axon. This can be reasonably assumed to represent a transport system from the RNA output in Nissl bodies to catecholamine and norepinephrine release from terminal vesicles at the ends of axons. These substances in the synaptic space associated with dendrites of associated neurons effect communication between neurons and within neuron associations. The holistic arrange-

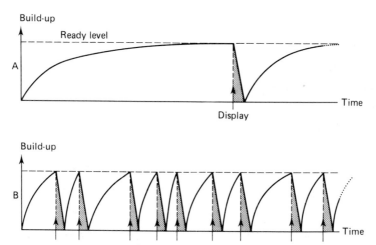

Fig. 2. A behaviour production model for tame domestic pigs, showing a low frequency of aggressive behaviour (A) as compared to that of wild domestic pigs (B).

ment ultimately generates behaviour. In a sense, neurons have a substantial secretory role not previously postulated. The secretory function requires, as in other glandular tissues and organs, phases of assimilation and dissimilation as the essential method of production of the glandular product. The production of innate behaviour is therefore subject to the characteristics of other bodily productions and appears to have closest resemblance to the synthesis of protein.

This concept is in accord with the Lorenzian account of the production of fixed action patterns (FAP) but is at variance with a common concept in which the output of behaviour is viewed as a product resulting directly from an input of stimulation. While this arrangement may still appear, superficially, to be so, it must be appreciated that stimulus input does not cause the production of innate behaviour but merely triggers the release of such relevant behaviour potential as has been adequately synthesized in the appropriate neuron circuitry.

4 Factors Affecting the Production of Behaviour

VISION AND BEHAVIOUR

Visual organization is a process which has been called 'an effort after meaning' (EAM), a process which undoubtedly plays a major role in animal behaviour. This 'EAM phenomenon' evidently resides at a high mental level; formerly it was thought to be essentially human, but this view is anthropocentric and erroneous. The EAM phenomenon is present and functional in domesticated and other mammals. It is personalized and exclusively directed in them towards resolution by close attention to immediate issues which are problematic to the individual animal. Altruistic communication to conspecifics of the interpreted meaning, when it is perceived, is a common and immediate response in animals by the use of vocalization.

The periphery of the retina serves to some extent as a warning system. To modify this function the eyeball moves almost continuously. When it does not move, it gives a 'glazed' appearance as a result of an 'over-steady' state of the eye.

The eye possesses the capacity for considerable mobility. The fields of vision can be altered by mobility of the eye itself, movements of the head and other major bodily movements. The visual world cannot be perceived comprehensively, only as a succession of images which can be perceived and noted. Vision is, therefore, episodic in character.

Light is made up of energy particles of electromagnetic radiation which can have both wave-like and particle-like traits. At times one trait predominates, at other times the other. The two properties operate in complementary fashion to give the comprehensive nature of light as a stimulus. Shifting from bright to dim light, or *vice versa*, involves a latency of visual adaptation of several minutes duration. The latency is longer in the shift from light to dark. The latency of adaptation is required for the essential process of depletion and restoration of the photoreceptors' supply of visual pigment (rhodopsin). This is a process known as bleaching and regeneration.

The emergence of variegated colours appears to result from a process in the visual pathway. This process compares the lightness of the separate images on wave bands provided by the 'retinex' (retina + visual cortex) system. The comparison of lightnesses of regions determines colour. The visual receptors are rods and cones. Peripheral vision is largely rod vision. Cones are rounded photoreceptor cells in the retina which provide both colour sensitivity and detail. The lightness code in the retinex system creates colour perception.

Ethological studies on perception of objects in highly motivated domestic animals allow us to recognize the fact that some of these animals, including ruminants, can observe colours. Recent behavioural studies on the mututal recognition of ewes and lambs, which have been artificially coloured, have given further proof of this fact. It must be appreciated that, in animal colour vision, appropriate motivation is necessary for satisfactory perception.

PHEROMONES AND BEHAVIOUR

Many animals communicate by means of their noses and by secretion. Basic to this mode of communication is a source of odour (usually a skin gland), a signal of odour (usually a secretion containing complex molecules) and the olfactory mucous membrane of the nasal cavity as a recipient organ. Further components are the stimulated centre, which is the olfactory part of the telencephalon, the olfactory sphere of consciousness located in the cerebral cortex and, eventually, the form of behaviour reactively produced. There is a tremendous variety among the odour-producing glands and organs of various species and also among the chemical nature of their products and their relationship to other organs. In competing for their large 'biotopes' ruminants have been quite well adapted to what the respective vegetation offers them and have developed a very effective social structure. Ruminants owe their success to their highly developed digestive system and, maybe more so, to their versatile system of olfactory communication. It is assumed that the odorous substances of mammals contain several pheromones at one time. Lately it has become known that mammals excrete pheromones not only through their skin organs, but also through the urine. One must distinguish between 'releaser' pheromones, which produce an immediate behavioural reaction, and 'primer' pheromones, which produce a retarded but sustained behavioural response.

Jacobson's organ is an additional olfactory receiver which is connected into the roof of the mouth and the nasal cavity respectively, and is instrumental in eliciting a reflex act which is known as *Flehmen*. *Flehmen* is a visible effect of the olfactory urine test.

The skin, as the primary protector of the body surface, is covered with glands serving temperature control, excretion, greasing and maintenance of the pH in fending off micro-organisms. This, in its totality, produces a specific body odour. In ungulates, in the convoluted type of gland in which apocrine excretion dominates, there is produced a volatile product which is moved to the skin's surface and mixed in a product which is particularly suited to contain, or bond, odorous substances.

To classify pheromones, one may say that they serve territorial marking, ranking, tracing, alerting, recognition and sexual stimulation.

HORMONE-DEPENDENT NEURAL FUNCTION AND BEHAVIOUR

Modern neuroscience has provided information on the ways in which hormones influence neural function when they attain threshold level in the presence of appropriate neuron populations. By this means hormones have a major influence on behaviour. Much behaviour is hormone-dependent; reproductive behaviour provides prominent examples of this. Some hormones liberate behaviour suitable for 'new role' situations.

With the recognition that there is chemical specificity among neurons, one can recognize that there must exist neuronal circuits which have the potential to produce but which may not be in regular use if the appropriate chemical message is not provided. The principal chemical messengers are, of course, hormones.

It would appear that hormones act on nerve tissue within the nerve system specifically to energize the production of neuronal output in these neuronal systems which are programmed to function in the presence of the chemistry of that given hormone. Since many 'new role' circumstances operate through the production of hormones, it can be seen that, in all probability, the purpose of such hormones is to activate neuronal systems which are not involved in routine maintenance behaviour. These new roles, or energy roles of behaviour, must nevertheless be established in the neuromechanism of the individual

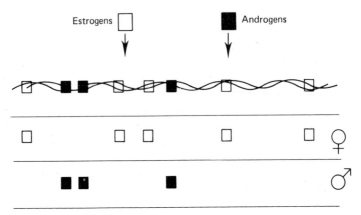

Fig. 3. Estrogens engage genetic code sections different from those engaged by androgens, resulting in different synthesis (production) by the neuron.

animal, to be called upon only when circumstances require their activation. In some circumstances their activation may never be necessary in the entire life of the animal.

It will be obvious that the neuronal circuits must be laid down in embryonic stages of development. It will also be obvious that, because of this, the circuits will be present and the same in individuals of both sexes. For this reason the sex hormones will activate only those neuronal circuits appropriate to their chemistry. This implies that neuronal circuits for the alternative sexual behaviour exist, but are not normally called into play in the lifetime of the animal. When abnormal circumstances occur, e.g. the presence of an ovarian tumour, the alternative neuronal system of sexual behaviour may then be motivated. This will result in the manifestations of behaviour patterns of the opposite sex.

The neuronal circuit for sexual behaviour of the alternative gender evidently does not lose potential merely because it is never called into play during the lifetime of the animal. The explanation for this is that the lifetime of the neuron equals the lifetime of the individual animal. Experimentally induced masculine behaviour in adult and aged female animals shows that exogenous hormones, of the inappropriate sex, are capable of producing the sexual behaviour controlled by the normally dormant neuronal circuit of that sex. Endocrine aberrations can be seen, therefore, as the basis of variable reproductive behaviour in animals.

Once hormonal behaviour has been potentially produced, this production will continue in the presence of the appropriate chemistry. This then places the production of behaviour potential, in this category, in the same circumstances as the behaviour of maintenance. As long as hormonally primed behaviour is being potentially produced, it will come under the same laws of stimulation and production as obtained in the case of behaviour patterns related to biological maintenance of a constant nature and which operate in a fairly constant state of chemistry.

It may be necessary to reflect on the subdivision of behaviour into that which is produced in the normally extant chemistry of the body and that which is produced only in the presence of short-lived chemical agents in the form of hormones.

5 *Acquired Behaviour*

Animals acquire many behavioural features through learning. Such learning helps to create the 'cortical mosaic' of continuing behaviour. It is important to appreciate the ways in which learning processes normally take place in farm animals. Learning adds to the repertoire of an animal's inherent behaviour. If its learning is deficient, an animal will remain deprived of some of the functions possessed by others of its kind and as a result it is likely to be a less adaptable individual than it should be. A study of the many factors capable of affecting the behaviour of an animal must include the various early experiences which can permanently affect the behaviour of that individual, even into its adult life. Environmental factors and forces have a much more powerful and durable influence when applied in early life than similar ones experienced in adult life.

The immature animal is more susceptible to learning than is the adult animal. Social and traumatic experiences tend to have more effect the earlier in life they are experienced. Animals are seen to benefit from as great a variety of environmental stimulations as possible in early developmental life.

An infantile store of experiences is accumulated from environmental effects. During the paediatric period the total effects of learning are compounded. The development of emotions, the opportunity to pursue exploratory behaviour, the social experiences of the young animal and the development of its physical and physiological apparatus, all combine to influence the animal's reservoir of acquired behaviour. Post-weaning environmental experiences also play their part in developing the behaviour of the animals in subsequent adult life. Learning processes continue into this period and senses also continue to develop

Fig. 4. Contactual nipping in the young foal, being reinforced and forming a habit.

Fig. 5. A young donkey foal being socialized with a food reward.

to improve the animal's awareness of its environment. Social experiences are still occurring at this mature age. Investigatory activities, from which much learning is derived, continue throughout life, though they tend to be more obvious in the young animal.

Several neurotransmitters have been implicated in learning, as in many other behavioural processes. It is suggested that no single neurotransmitter subserves one behaviour but a prime neurotransmitter in learning is evidently acetylcholine. The number of vesicles of neurotransmitter substance in neural terminals apparently increases with learning. Neural conduction rates increase with learning and it seems that sensitivity to neurotransmission increases with learning.

MODES OF LEARNING

The main effects of learning in domesticated animal behaviour are seen in modification of species-specific behaviour. In this way innate behaviour becomes functionally important and shaped to circumstances. Learning, therefore, conditions innate behaviour in animals. Such learning develops through various processes. Two principal processes of learning are classical conditioning and operant conditioning.

Classical Conditioning

In classical conditioning, which is the Pavlovian form, a behaviour stimulus (unconditioned), such as food availability, becomes closely paired with another (conditioned) stimulus, such as sound of food preparation, with sufficient frequency that the conditioned stimulus alone triggers the behavioural response. Examples of classical conditioning abound in domesticated animal behaviour. Pigs become roused when feeding sounds commence; cows assemble at feeding troughs when the sounds of feeding equipment or personnel are detected; stud horses become activated when led out by their grooms for breeding; sheep approach a trailer carrying winter fodder as soon as it appears.

Classical conditioning is a form of learning which typically speeds up responses. The unconditioned stimulus acts as a primary reinforcer to the behavioural response. The conditioned stimulus is, therefore, a secondary reinforcer. Reinforcement is the result of the reward or the return obtained by the responses. Some stimuli achieve positive responses, some trigger negative activities. For this reason, it is sometimes convenient to recognize positive and negative reinforcers which may, in their own turn, be unconditioned or conditioned, i.e. primary or secondary. The examples of negative reinforcement would include most avoidance behaviour such as the avoidance of persons, or the appearance of persons, associated with aversive stimulation. In the following examples the reward is avoidance of undesired experiences. Horses may run from the approach of a handler as the result of negative reinforcement. Sheep bolt from human approach as a result of similar learning. It is good behaviour management, therefore, to reduce painful and frightening incidents, particularly in the young animal's first experiences of significant events such as control, treatment and handling. Classical conditioning supplies many learned refinements to the behaviours of reactivity, in the class of maintenance behaviour.

Operant Conditioning

The second major type of learning is instrumental or operant conditioning. It is, in effect, trial-and-error learning and is the learning which occurs from the numerous empirical activities which are generated by exploratory and investigative behaviour. Animals are very dependent on operant conditioning for learning performances of behaviour that do not constitute natural innate actions. It is likely that in some training procedures circus animals' tricks are taught before feeding and by coaxing the animal to operate within the circus environment in the desired ways. When appropriate behaviour is shown, a food reward, or reinforcement, is given. Reinforcement leads to improvements in learning and to the establishment of new behaviours. As learning proceeds in operant conditioning the animal makes fewer empirical actions, until eventually it performs proficiently. Operant conditioning is also called instrumental learning or conditioning, since the behaviour is the instrument by which the reward, or reinforcement, is obtained. Training animals is an operant task, the trainer may wait until the animal produces the desired activity. This is then rewarded promptly, as a reinforcement. This process of learning can be speeded up by 'shaping' the behaviour.

Training Principles

In training, behaviour is typically 'shaped' by initially rewarding the generic action and subsequently presenting the reward only on the production of more specific actions. Repetition is important, as is consistency, and in this respect animal training resembles drilling. The standard training procedures are analogous to 'shaping' in experimental animal psychology. A start is made by rewarding each successive activity which approximates the behaviour which is desired. Therefore reward is given only to those behaviours which are closer to the objective of the training.

Many circus horses are said to be 'shaped' by being rewarded for fairly natural activity such as trotting, in a circus ring (Fig. 6). The horse may be further rewarded for rearing on a command. The command is likely to be associated with the flourish of a whip to encourage the rearing attempt. Eventually, the trained horse can be induced to rear and walk on its hind legs. Animals that are trained to perform complicated and relatively unnatural tricks are usually continuously reinforced with food rewards. The main characteristic of this type of learning is that the animal's behaviour is manipulated and under constant discipline.

The skill of the trainer lies in recognizing small progressive responses and rewarding each of these. Even the smallest progress in training may be the key to the desired performance. As training becomes completed the desired responses alone will be reinforced by reward. In training which involves an elimination of undesired behaviour, such as biting, the conditioned painful response must be paired very closely in time

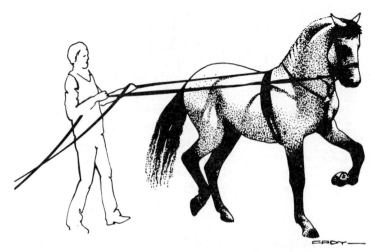

Fig. 6. 'Shaping' the way-of-going in a performance horse.

with the misbehaviour. For instance, horses can usually be trained to stop biting by being pricked with a sharp nail or pin in the upper lip at the instant of an attempted bite. If punishment is to be used for misbehaviour it must come as soon as possible, preferably within seconds, after the animal's offence. The punishment conditions and the avoidance of pain reinforces the modified behaviour. It cannot be overemphasized that punishment must be very closely paired in time with the misbehaviour if the animal is to learn appropriately.

Other types of learning occur which do not fall into these classes above. Examples include mimesis or imitation and habituation. A notable form of animal learning is that which occurs very quickly and easily during critical periods.

Critical or Sensitive Periods and Imprinting

A critical period is a time in an animal's life during which an important behavioural development can be facilitated. If the development does not occur during that time it may never occur. It is this feature, therefore, which is critical. Critical periods have also been termed 'sensitive periods'. Some of them coincide with episodes of acute operation of the senses, such as sight, hearing and smell. During a critical period the

Fig. 7. A newly calved cow commencing care-giving during the sensitive period.

animal is particularly susceptible to fast learning, involving specific cognitive ability. The animal acquires durable recognition of, and affinity with, specific environmental content in such periods of phenomenal awareness. Animals in critical periods may be at a given stage of neural development or in very special hormonal stages.

The best example of learning in a critical period is imprinting. Imprinting is the rapid formation, during the early post-natal period, of

Fig. 8. Rapid bond formation between mare and foal in the postpartum critical period.

a permanent close attachment between an animal and a salient environmental object, such as its mother. Imprinting, as a phenomenon, is a combination of learned and instinctive behaviour. In imprinting, the animal's genetic programming causes it to be maximally sensitive, at the critical period, to a moving object, so that it can speedily learn the strong habit of following a specific object. This strong following habit does not seem to be built up over a duration of time or with practice. The animal's following behaviour is apparently learned completely in one episode.

With 'instant learning' in critical periods, it is as though an area of cerebral cognition is presented as a *tabula rasa*, or clean slate, ready for inscription. The objects to which the animal in the critical period is particularly susceptible are also those to which the animal is oriented with a high degree of motivation. This latter state apparently helps to establish the animal's special learning ability. In general it appears that proficient learning in animals is often achieved during states of appropriately oriented motivation.

A critical time occurs in the puerperal period of the mother, complementing, in many cases, the time when imprinting is operative in the young. This period is variable in duration but is measurable in hours in those ruminant species in which it has been studied, such as goats, sheep and cattle. At this time the mother quickly acquires, or learns, the identity of her own and thereafter relates to it like a particularly vulnerable extension of herself. Care and protection are provided to the young, acquired as property during this period (Fig. 8). In contrast, other young, even if evidently similar, are rejected; hostile behaviour is exhibited to the alien. Such behaviour may amount to extreme aggression in many cases. When the maternal critical period is passed, adoption of fostered young by learning is more protracted and less certain.

Other critical periods occur and these are typically at key biological points, such as puberty when sexual orientation may be enhanced by learning. It is possible that several other critical periods in animal behaviour await discovery.

Learning by Observation

Animals observe one another closely in a great variety of circumstances. Learning by observation is a basic technique and takes place commonly in animals. Animals watching others act and react, subsequently adopt similar behaviours. Much learning in young animals is by observation and they learn more readily by watching their mothers than by watching other adults (Fig. 9). Their mothers, of course, permit close investigation by their own young. Grazing animals, again particularly young ones, learn from others such things as food selection and location, paths and routes, watering places and shelters. Needless to say such learning is of critical importance in the ability of the young animal to integrate successfully with its environment and home range. The role of such natural learning in survival can be seen to be considerable.

Fig. 9. Foal learning the eating habits of the mare.

LEARNING AND MEMORY

Learning forms memory. This appears to take place by an initial formation of a short-term memory deposit and the subsequent formation of a long-term memory account. The latter consolidates learning behaviour. The formation of long-term memory traces appears to require variable periods of time in those animals in which this phenomenon has been studied by task learning, but the two stages of learning formation are readily discernible. Memory formation eventually involves intraneuronal protein synthesis, in the same manner in which the production of behaviour potential in neuron circuits is basically a manifestation of protein synthesis. Various evidence exists to show the role of protein in memory. It has been shown experimentally that isolation and synthesis of one protein, scotophobin, causes untrained mice to learn a specific task following the injection of this protein. Other evidence is accumulating, in biological studies, showing that specific proteins are involved in specific learning.

The bulk of the cerebral cortex contains associative areas involved in memory, but the existence of a 'learning centre' in the brain remains unknown and is doubted. There are some forms of learning, however, which do not require cortical function. A principle of mass cortical action, however, appears to operate in comprehensive memory formation. Learning may be affected by brain lesions. Learning and memory of a previously learned task are subject to loss with brain damage. But not all lesions impair learning. For example, lesions in the ventromedial hypothalamus have actually improved learning of conditioned avoidance responses in experimental animals. With brain damage, the more cortex that is lost the more severe are the deficits in learning and memory that occur. Older memory traces are evidently the most consolidated and are resistant to removal. This indicates that well established memory traces may be diffusely deployed throughout the cerebral cortex.

INTELLIGENCE

Numerous enquiries were made in the early part of this century into the comparative intelligence of animals. In these studies attempts were made to make facile comprehensive appraisals of animal intelligence. The implicit assumption was that animals and their behaviour could be better understood through proper comprehension of their intelligence. The assumption was faulty. Although intelligence in mankind is well understood and in many cases affects human behaviour, it cannot be assumed that intelligence in animals is of an identical nature. It can now be recognized that intelligence in animals is more general than specific and that its role in behaviour is apparently a supportive one based on essential sensory perception. Nevertheless animal intelligence does exist. It is technically termed noesis, which is defined as 'the sum total of the mental action of a rational animal'. But noesis is not a critically important factor in behaviour, although species-specific properties of intellect can be found in animal behaviour. While various methods are often used to attempt to measure intelligence in animals—although this is not a profitable or practicable exercise—it would appear that the length of time that an animal can remember a specific signal of training or command can be taken as some measure of intelligence. Promptness in learning is also another evidence of intelligence and many examples of this can be seen. Animals learn many techniques in maintaining themselves in domestication. Pigs use operant conditioning very readily to make optimum use of their environments. Very young pigs, for example, readily learn to operate self-feeders with hinged covers and automatic waterers, using their

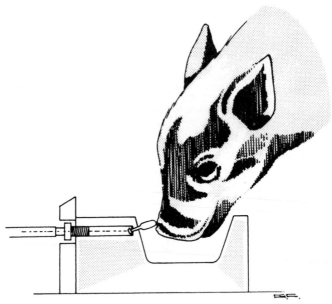

Fig. 10. Operant conditioning in a piglet which has learned to push a lever to obtain milk substitute.

snouts (Fig. 10). All farm animals readily learn to operate automatic watering devices. Intelligent traits are under some genetic control. Some species learn with greater facility than others. Some breeds within species have characteristic differences in learning ability. Variation in intelligence between breeds and within breeds is noticeable.

6 Environmental Influences on Behaviour

CLIMATE

The adaptive mechanisms of cattle, living under various climatic conditions, have come under increasingly detailed study by animal scientists over the last two decades, but it is only recently that the behavioural aspects have been closely examined. The differences in response to heat, solar radiation and the immediate environment are thus becoming better known and they are found to vary greatly between breeds and from one area to another.

The level of temperature at which the so-called European breeds of cattle are able to maintain a normal body temperature (the thermo-neutral zone) is said to be about 0° to 20°C. Tropical breeds, on the other hand, are able to maintain a normal body temperature in ambient temperatures of about 22 to 37°C. There is evidence to suggest that some tropical breeds are even able to carry on normal activity and locomotion in temperatures in excess of 37°C. Both types are, however, found to use behavioural methods in attempting to control their temperatures.

Thermoregulation

The first and most easily recognizable evidences of adaptive behaviour in cattle are the movements directed towards seeking shade, particularly when ambient heat greatly exceeds body heat. The fact that European cattle show a stronger motivation towards shade-seeking, and have a greater shade-dependence, than tropical or subtropical breeds, is not unexpected. Thus, though the behaviour of both Aberdeen Angus and Brahman cattle in cloudy, over-cast conditions is much the same, when the two breeds are in conditions of direct solar radiation, with little or no air movement, differences become apparent. There is a definite change in the behaviour of Aberdeen Angus, which seek out shade and consequently spend less time grazing than do Brahman cattle, which are less inclined to undertake adaptive behaviour under such conditions. On the other hand, the adaptive behaviour of the Aberdeen Angus is much less marked if there is good air movement, regardless of the change in temperature. Cattle living in tropical rain forest and equatorial areas of the world show a greater need for shade than those living in sparse or semi-arid areas where rainfall is low and shade limited. It has been found that Dwarf Shorthorn cattle in Nigeria may spend as long as 4·5 hours resting in the shade during the day and, to compensate, as long as 3·5 hours grazing during the night. In this way their behaviour may be nearer that of the temperate breeds, such as Aberdeen Angus, Hereford and Holstein, than that of the Brahman breed.

Ingestion

An early behavioural change during high temperatures is a reduction in food intake. This may arise not only from the shade-seeking activities of the animal, but also from a reduced desire for food. Consequent results are an alteration in eliminative behaviour and a general decline in performance.

In subtropical conditions cattle often reduce their water intake to once every three days. In areas with low rainfall and a long dry season, cattle normally graze up to 6–8 km from the nearest watering place, but a proximity to water of 19 km may be enough to maintain a satisfactory degree of health and survival. Often the entire grazing area around the water is completely eaten out. In such cases it is the cattle which are strongest and the most resourceful foragers, which can expect to continue as viable livestock. As the dry season approaches, there is a noticeable reduction in water intake and if breeds of tropical cattle are deprived of water the nature of their metabolism enables them to put on fat. They are able to survive in this manner for up to two months, until the fat reserve is reabsorbed. By virtue of their long intestines (much longer than those of European breeds), cattle of tropical origin are able to reduce the volume of urine and faeces, thereby conserving much of the liquid contained therein. It is this character which enables these animals to travel on foot for hundreds of kilometres to grazing areas, with the aid of experienced drovers, and to arrive there alive.

Breeding

In the tropics sexual behaviour is affected by seasonal conditions. Young animals have a much reduced chance of survival if parturition has taken place during seasonal weather of an extreme nature.

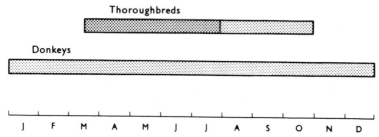

Fig. 11. The span of the breeding season in donkeys and Thoroughbred horses in a tropical latitude (18°N) in terms of the spread of estrus. The presence of a seasonal feature of breeding in horses and the existence of a more intensive breeding phase at the start of annual breeding activity are shown.

Consequently natural reproductive activities occur more often during certain periods of the year than during others. Generally, the sex drive of the male is found to be least apparent during the hours of greatest solar radiation; in the case of sheep, temporary impotence often occurs. In certain areas such as Africa, where there is only one rainy season per year, breeding is largely restricted to that period.

Observations on the reproductive behaviour of male goats in the West Indies have shown that these animals are able to mate throughout the year regardless of seasonal changes and that the level of the libido remains fairly constant. It was found, however, that sexual behaviour was depressed during wet weather. It is an aspect of these goats' general reproductive behaviour that they are least active during rainy periods.

Boars are seen to have little or no interest in estrous sows during particularly hot weather and a similar lack of interest can be seen in bulls. If the animal's body is wetted or it has facilities for wallowing, however, the level of libido returns to normal. In general the level of libido in the male is reduced under conditions of intense heat.

On the other hand, a drop in temperature raises the level of sex drive in some animals and can even initiate breeding in those species which

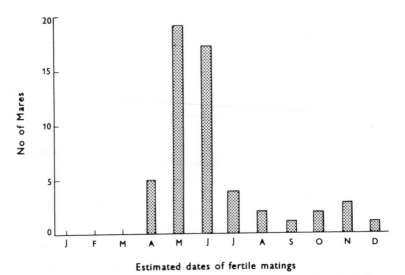

Fig. 12. Fertile matings in a herd of free-ranging Exmoor ponies. The full span of the breeding season is revealed only by the less fertile mares.

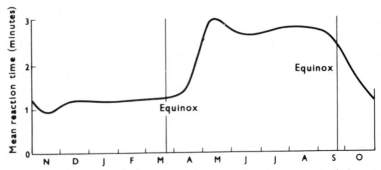

Fig. 13. The seasonal trends in sex drive in nine goats as shown in their mean reaction times. Two principal phases are recognizable as plateaux in winter and summer. Note the two short periods of high and low sex drive after autumnal and vernal equinoxes respectively.

exhibit strong sexual behaviour seasonally. This attitude has been seen in sheep in central Europe, but it is rarer and less marked than their reaction to heat. The reaction of the male to colder weather is usually observed in seasonal breeding mammals, but this is partly obscured by the fact that cold weather generally has an adverse effect on estrus in the female and hot weather a slightly favourable one.

Among those species in which the breeding behaviour is not specifically restricted to any particular season of the year, there are definite occasions within the 24-hour cycle when the sex drive is more apparent than at other times. Many breeds of sheep, which maintain an all-year sex drive, are known to mate more often around sunset and sunrise, particularly the latter. As the onset of estrus in ewes usually takes place at night, the period around dawn has the highest incidence of mating. It has been verified in Brazil, among Merino, Romney Marsh and Corriedale sheep, that estrus occurred at night in 75% of cases and a similar pattern has been observed in Cheviots in Britain.

Probably the main environmental factor influencing sexual behaviour in seasonal breeders, such as horses and sheep, is the ratio of light and dark during the 24-hour cycles throughout these seasons. The photoperiodic influence, however, is a complex one, combining the intensity or quality of light with the relative duration of light and darkness. The positive influence of photoperiodism on the animal's sexual behaviour is usually associated with an increasing light-to-dark ratio. Some species, however, show a corresponding escalation in sexual behaviour when this ratio is reversed.

Experimentation and observation have shown that the manipulation of the photoperiodic influence on reproductive behaviour is often possible by artificially advancing the season. An earlier onset of estrus caused in this way usually levels out as the season progresses. One aspect of the manipulation of the day length is that it can advance the development of puberty in the young female or, alternatively, slow it down. A reduction in the hours of light has successfully induced photoperiodic and behavioural responses bringing about estrus. On the other hand, a progressive daily increase in the hours of darkness has successfully brought estrus to an end. It has been found that a progressive decrease in the amount of daylight is not the only successful way of encouraging estrus; a simple fixed ratio of light and darkness has also proved successful. There is also evidence that not only light but also an unbroken duration of darkness influences estrus and that patterns in the duration of darkness can bring forward the beginning of the breeding season in sheep. Mares can be made to come into estrus by exposure to prolonged periods of light out of season.

Less research has been done on male sexual behaviour and the degree to which it is subject to seasonal variations. It is often assumed that the male will always be ready to serve at the time the female comes into estrus. Observations were made in Canada on the breeding rhythms of male Saanen and Toggenberg goats in natural light conditions. It was found during the period of prolonged daylight lasting from around April until September, sexual activity and potency were relatively weaker, but that the reaction times returned to normal in October. Thus it was deduced that there is not so much a total inhibition of the sex drive in summer, but rather a comparative reduction, whereby the drive is more easily satisfied and has longer reaction times.

TRANSIT BEHAVIOUR

Cattle become cooperative with handlers and with each other when travelling long distances by rail. When travelling together in an enclosed area, they rarely move about to alter their position or seek another part of the wagon and do so only when the train has stopped or is stopping or starting. They travel facing the side of the wagon and with their bodies at right angles to the direction of travel. This appears to be the most convenient position when seeking optimum comfort and space for balance and for minimizing injury should they lose their balance. It has been found that cattle with longish horns usually adopt a position where their heads are resting upon the backs of the adjacent animals; this makes for easier breathing as well as avoiding injury.

On long journeys it is often over 24 hours before some of the animals seek rest by lying down, but others avoid doing so for the entire journey. They usually lie down in groups of three or four after one animal has made the initial move and tend to do so at one end of the wagon. The remainder of the group will avoid trampling these animals as much as is possible, quickly moving any hoof which is brought down on the body of a recumbent animal. When given the opportunity to rest while disembarked at yards in sidings, they move to the water trough or walk about for a very short period, even when food is clearly visible, and feed only after their thirsts have been quenched.

At the beginning of the journey urination and defaecation in the wagon are frequent, but decline as the animals adjust. Cattle which have had neither food nor drink during a long journey sometimes expel small dry pellets of dung and may still urinate.

The hazards to which animals being transported by sea are vulnerable are numerous. When entering equatorial regions cattle are subject to heat stress. There is a constant danger of injury in rough weather. Horses need plenty of head room at sea. Cattle are highly susceptible to injury and illness during sea voyages.

SOCIAL DENSITY

One aspect of the behaviour of animals in close proximity is the strong allelomimetic tendencies which they display. This behaviour, whereby one individual is influenced by the grazing and drinking activities, eliminative behaviour or movement of others, is seen in a clearer light when the animal is placed in isolation, out of sight of any of its herd (or flock) mates and confined there. The animal shows visible signs of stress and its habit patterns become disrupted. Its intake of food and water declines and it makes constant efforts to rejoin the main group if it can possibly do so. Some sheep and horses have in fact been shown to decline all food and drink when totally isolated, even though both food and water may be immediately available. The isolation of an animal from its herd should therefore be avoided whenever possible, as the effects of this treatment on the animal can often be as undesirable as those of overcrowding.

Excessive social density in animals frequently results in restlessness, fighting and a disruption of daily routines of behaviour. All members of the group seem to be similarly affected, whether they are directly involved in the agonistic episodes or not.

The effects of group pressure on behaviour are various, but in general it appears that social concentration potentiates some primary drives. Within groups, the drive of the majority seems to prevail so as to direct behavioural policy for all. A notion of social superdrive comes to mind to serve as a concept of the holistic strategies of group behaviour. Social superdrives, in flocks and herds, may be motivating forces in stampedes, marches and migrations which persist as outstanding behavioural phenomena in animals. The social drives have the means to derive increased motive force from (1) allelomimetic pressure, (2) suppression of intra-species aggression, (3) cumulation of individuals' drives, (4) unified reactivity and (5) synergism of compatible primary drives such as those of association, kinesis and exploration. In addition, mass communication is apparently present, operating by methods as yet only partly determined, to arouse and intensify innate activities in a population. A superdrive is therefore a hypertrophied drive which has developed from group effect or social facilitation.

Population excesses are known to create ethological malfunction and, to acknowledge this further fact, the hypothesis might require extension as a tentative law of social concentration as follows: 'operative association in animal groups potentiates group drive harmony, while saturated association disturbs it.'

RESTRAINT

The use of direct force in controlling the behaviour of animals is best exemplified by the variety of 'crushes' or stocks that are in use on many farm premises for the tight restriction of movement in the larger farm animals. Crushes of various types are in operation for controlling cattle, for example, but the best forms of crushes allow an animal to be funnelled down a narrowing serpentine passage into the crush section. At the exit of the crush there should be a small collecting yard where animals which have already passed through the crush may be seen by the animal entering or within the crush. Crushing arrangements which make use of this broad principle allow large numbers of cattle to be examined individually and closely in a short space of time with the least amount of danger to themselves and to those handling them. Such crushing techniques allow mass treatment of herd or flock for operations such as vaccination, drenching, ear-tagging, blood sampling, tuberculin testing, pregnancy diagnosing, branding, spraying and dehorning. Many of these operations can cause animals to become so alarmed that they respond behaviourally in ways which frequently cause injury to themselves. Whilst restrained within stocks, cattle frequently attempt to escape by pushing forwards. They also frequently attempt to push their hind limbs against one side or other of the stocks. The stocks should therefore be constructed with solid sides so that it is impossible for an animal to put its hind leg between spars since, when this happens, serious injury to the limb is likely. After stocks have been in use by a number of animals they tend to become slippery underfoot with the excreta of these animals and it may be necessary to improve the footing within the stocks by the addition of ash or sand from time to time. When crushes possess some yoke arrangement which grips a restrained animal by the neck, it is important for there to be an efficient quick-release mechanism which allows the animal's neck to be released quickly should its limbs slip from under it and cause it to fall within the stock. This is particularly important when large animals such as bulls are being put into stocks. Their weight prevents them from being raised manually should they fall to the ground. Fixed crushes in which a string of animals can be restrained tightly behind one another, perhaps with intervening bars separating two or three of them, are known as races. This form of crush is very suitable for a close inspection of a large number of animals in a short time.

Stocks for horses are sometimes employed when an animal is to be examined per rectum, for example. Such stocks are also suitable means of restraint when some operation to the feet is being carried out on a fractious horse. Stock sizes vary with the type of horse to be examined. It is also essential that they should be extremely solid. A horse in close restraining stocks, which finds the stocks moving or hears the parts moving, is very likely to become overexcited and to lash out in a frenzy of kicking in a manner which is difficult to control.

It is principally with regard to cattle that the problem of handling of loose animals is greatest. In many forms of cattle husbandry, calves are already well grown before they need to be handled for the first time. Such calves are difficult to catch and eventually to control. A full-grown calf, unaccustomed to being handled, may show different forms of behaviour when approached. An initial state of alertness is usually observed. The animal directs its attention exclusively towards the source of the approaching danger and its behaviour reveals that it is conscious of the principal stimulant in its immediate environment. Its head is directed towards this source and its eyes, ears and tail are moved in ways indicating its total concern with the person approaching. Closer approach towards the calf usually induces a state of alarm or fear. A group of animals will quickly share this state of alarm by generating signals which are understood by all the others of the group. Close herding results: the animals pack more tightly together; they move more rapidly; their heads are held up; there is likely to be some bellowing. Animals in this state have been described as being fully adrenalized. In this condition they offer more resistance to handling. To control the behaviour of calves in this state, the approach by the handler towards the calf should be quiet, even in pace, but cautious. During this type of approach the calf will start to move away, either forwards or backwards. The handler should then modify his approach, either to the left or to the right, so as to cut off the intended route of escape. After one or two intention movements the animal will then direct its flight towards a corner. The expert handler makes use of this knowledge of animal behaviour to time his attempt at catching the animal to coincide with the time when the animal's head is directed towards a corner. When the calf is being caught it is important to grip its lower jaw and to raise it and pull it sharply to one side. Gripping the lower jaw is more effective in controlling cattle than catching the nose. Small calves in particular resent their noses being held. This part of the body is apparently very sensitive and the restriction to breathing also induces further panic.

In directing the movements of animals individually or in groups, the expert handler makes use of the fact that animals apparently regard as part of the approaching person anything which is held in the hand and extended outwards on either side. In this way one person can extend his presence considerably to either side and thereby cut off a very large escape area for an animal which is loose.

This general rule is not applicable to pigs. They do not appear to be susceptible to this form of control. Pigs are best blocked from their escape route by placing some solid object, such as a board, at ground level directly in the line of the escape. Once a pig has selected a line of escape it is not so likely to deviate from this line as are other animals and, when this route is blocked in the manner described, the pig is frequently stopped altogether in its flight.

Part II
The Behaviour of Development

7 Behavioural Development

Features of behaviour emerge and change during the development of the animal. The courses by which these changes occur have now become recognized as an important branch of ethology. This special area of behavioural ontogeny not only reveals crucial patterns of infant behaviour in farmed animals but also shows the phenomenon of a time dimension in some innate programming of behaviour. A full appreciation of the behavioural as well as the physical development in young livestock is of critical importance in their optimum management. Mortality rates in these animals are great and can be controlled only by progressive management, taking account of their specific behavioural needs. Development of behaviour in the fetus, the neonate, the nursling, the weanling and the pre-pubertal farm animal has received some keen study in recent years. Different species have provided excellent models for different aspects of this study; equine subjects have been particularly useful.

FETAL BEHAVIOUR

Fetal behaviour in farm animals is a subject of increasing interest. It has been studied by ultrasound, radiography, palpation and surgery in horses, cattle and sheep. The use of an ultrasonic approach has permitted the study of quantitative fetal movement. Fetal kinesis is the major form of fetal behaviour. It occurs throughout the greater part of pregnancy and, with gestational progress, different features of fetal kinesis become recognizable. Simple, singular movements are of common and quite regular occurrence from mid-gestation onwards.

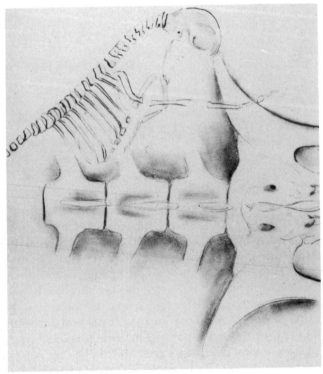

Fig. 14. Ovine fetus in mid-gestation showing extreme flexion of one forelimb and extreme extension of the other.

Such movement maintains a fairly constant daily incidence, averaging 54 per hour in the equine fetus and 33 per hour in the fetal calf.

Radiography has been used in sheep to observe certain qualitative aspects of fetal activity during days 80 to 120 of the 150-day gestation in this animal. Simple fetal movements observed by this method are generally similar in form and of characteristic occurrence in each of the major mobile anatomical parts. The neck, forelimbs and hindlimbs show most movement. These movements occur independently of fetal position. They are not subject to influence by the number of co-fetuses. Radiographic study has shown that simple fetal movements usually consist of exercise movements of major parts. These exercises are common and apparently autonomous. They take the form of extension and flexion, principally of the limbs and neck.

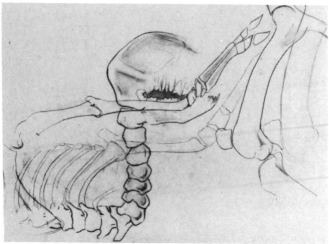

Fig. 15. Drawing of a radiograph illustrating the following components of puerperal 'righting' behaviour: carpal extension and head and neck elevation. Note that rotation of the thorax has not yet commenced. (Inlet of the maternal pelvis is top right.)

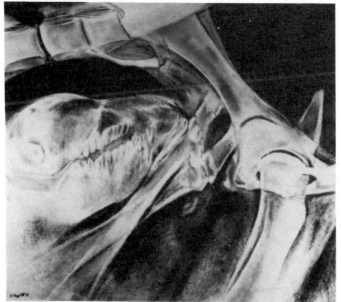

Fig. 16. Equine fetus showing extension and head and digits of the fore limbs as part of the 'arousal' phenomenon prior to parturition.

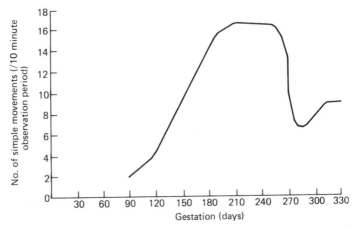

Fig. 17. Distribution frequency of means of simple fetal movements in 70 observations in 38 pony mares.

It is obvious that repetitive muscular activity in fetal kinesis will inevitably have the effect of establishing improved muscular development and muscle 'tone'. Evidently simple fetal movement is a phenomenon directly related to physical maturation, muscular development and kinetic competence.

Movements of greater complexity emerge in the kinetic behaviour of the more mature fetus. Coordination of complex fetal movements develops and leads to their grouping into phases. Mass fetal activity occurs in lengthy phases in the prepartum stage of gestation. The peak of fetal activity occurs three days prior to parturition in the equine subject and about two to three days prepartum in the bovine fetus. Mass kinesis, involving approximately 3500 movements, coincides with fetal 'righting' actions and typically ceases one or two days before birth, when a phase of quiescent fetal behaviour prevails. Evidently the muscular competence necessary for such a scale of fetal activity is acquired by intragestational fetal exercising. Attainment of the birth posture by the fetus is the apparent function of terminal fetal kinesis.

The posture of alerted behaviour has been found to be most proficiently secured by the heavier, more muscular, fetus. This finding indicates the need for adequate muscular development if the fetus is to respond properly to terminal arousal, as yet an undefined phenomenon. Postural defects, chiefly head deflexion and forelimb flexion, appear to

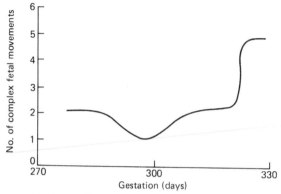

Fig. 18. Sequential observations on complex fetal movement in three mares in the last 60 days of gestation.

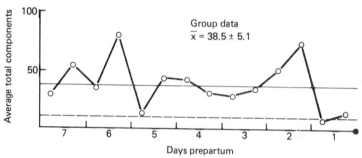

Fig. 19. The course of kinetic activity in a group of 12 bovine fetuses in the last week of synchronized pregnancies.

be associated with aberrant kinetic patterns or relative fetal inertia at full term when 'alerted' fetal behaviour should occur, about three days prepartum in cattle. After the alerted stage, the fetus becomes involved in more mass and complex movements which collectively create the 'righting reflexes'. The first component act is carpal extension, which directs the forelimbs towards the maternal pelvis. The head and neck are then raised and outstretched in the direction of the pelvic inlet. Later changes in fetal posture involve rotation of the trunk. Essentially, these righting reflexes dictate a conversion from general flexion to strategic extension of certain parts; they evidently occur during the two days before birth in the sheep, one day prepartum in the bovine fetus and in first-stage labour in the equine subject. Further specific postural features of note during birth include the flexion and retraction of the elbows against the thorax and terminal conversion of the hind limbs from full flexion to full extension.

Certain general features of fetal kinesis have become evident. Fetal posture and engagement at the pelvic inlet become fully organized on

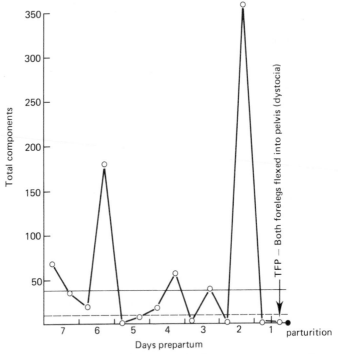

Fig. 20. Abnormal pattern of fetal kinetics terminating in abnormal terminal fetal posture (TFP) in a bovine subject.

the last day of gestation, sometimes only in the last hours in cattle. Absolute inertia of the fetus on the last day is likely to result in fetal malposture and birth difficulty. Inertia follows abnormally high spikes of fetal activity and is apparently a form of fatigue in the fetus. Episodes of massive fetal kinesis precede the alerted phase when fetal postural extension first occurs, usually a few days before pelvic engagement and birth.

The studies on fetal behaviour have drawn attention to the manifold role of kinesis. Attention is especially drawn to the quantity of kinesis contained in the prepartum period when righting behaviour is creating the birth posture. The work required in this could not be accomplished without a state of muscular fitness. It is an obvious conclusion that continuing fetal activity during gestation must have this purpose. The biological function of fetal kinesis is apparently two-fold: firstly to establish muscular competence and secondly to create the characteristic birth posture. Together, these functions facilitate intrapartum survival, doubtless the ultimate objective in fetal kinesis.

The details of all the coordinated ungulate fetal behaviour constitute evidence of critical ethological competence in the normal mature fetus. This evidence is sufficient to undermine many conventional concepts in which the start of mammalian ethological function is timed from birth.

8   General Postnatal Behaviour

The development of locomotion is telescoped into the first few hours following birth in all the farm animals. Neonatal behaviour contains several major initial features. The day-old calf, lamb, foal and kid all show bursts of sudden capricious behaviour as spontaneous acts of locomotor play in the form of leaping. In calves this leaping activity takes the form of a little dance. The neonatal bond to the mother is a major ethological phenomenon. The neonatal work involved in this bond formation falls into several categories.

Dependent upon prior fetal competence, together with favourable puerperal circumstances, satisfactory neonatal survival requires great adaptive success. The latter is dependent on competence in the bonding of the newborn individual with its dam. Seven primary and co-essential stages can be recognized in the comprehensive repertoire of ungulate behaviour during the formation of the neonatal–maternal bond. These are coordinating recumbency, elevation, ambulation, environmental exploration, orientation, udder searching and ingestion.

ESSENTIAL STAGES IN POSTNATAL BEHAVIOUR

Coordinating Recumbency

Immediately following its expulsion at birth, the neonate lies in extension, raising the head and neck. The animal then completes rotation on to its sternum and 'collects' its hind quarters. The head is shaken and the limp ears become mobile. The forelegs become flexed. The head is carried to the inside of the long spinal arc. Whilst still recumbent, the animal is then balanced and in a position from which it

Fig. 21. First postnatal activities in a foal, showing elevation of the head, extension of the forelimbs and mobility of the ears.

can attempt to rise. This recumbent stage facilitates grooming by the dam over dorsal areas of the neonate.

Elevation

In a manner typical of the species, the neonate ungulate attempts to rise to an erect stance as the second stage of postnatal behaviour development. The antigravity function of the vestibular system is apparent in this behaviour. Typically, more than one attempt is made to establish upright equilibrium. It is common for a 'half-up' posture to be struck and held as a first measure of successful rising (Figs 22, 23). Extension and muscular tension of the forelimbs are initial features of the upright stance.

Ambulation

When the stance has been secured, attempts at walking are subsequently begun. First steps are the typical four-step form of slow ambulation or

Fig. 22. The half-up posture of the foal as success is gained in rising attempts.

tentative walking. Unsteadiness is a prominent feature and apparently exaggerated, perhaps advantageously, by the presence of the eponychia, or fetal digital pads. These have a protective role during fetal behaviour and are still present on the plantar hoof surfaces during the first walking activities. This eponychial tissue rapidly becomes shredded and removed from the soles by wear during early ambulation. The unsteady form of locomotion normally secures further maternal attention. Neonatal motility in general stimulates the maternal drive.

Environmental Exploration

Myopia is a constant neonatal handicap, the limited visual competence evidently varying in degree between species. Exploratory activities typically take the form of 'nosing' environmental features with the head fully extended on the same level as the trunk. In these activities the neonate may be intercepted by strategic positioning of the dam.

Orientation

The dam is normally encountered in early exploratory exercises. She then becomes the principal focus of neonatal attention and orientation

Fig. 23. Foal standing for the first time and directing its attention to the inanimate environment.

as imprinting develops. In many ungulates there is considerable circumstantial evidence that the shaded underbelly of the dam serves as a releaser, providing the neonate with its only initial visual cue. The neonate is certainly drawn to the shadowed and darkened underline of the mother. Close examination of the limbs and ventral contours of the dam is performed by nasal touch. Tactile nursling hairs, now anatomically identified on the neonate muzzle, evidently function in this context. Orientation towards the dam is facilitated by discrete postural and positional adjustments on the part of the latter.

Udder Searching

Thigmotaxis undoubtedly influences the behaviour of the neonate in relation to its dam. The orientation towards the dam increasingly becomes directed to the mammary region. Mammary localization is seemingly determined empirically, but systematically, by trial-and-error examination of the ventral abdominal curvature. The udder is apparently identified by its tactile characteristics. The protruding teat quickly becomes the precise focal point in attempts at oral prehension. Successful teat-seeking becomes learned and is often aided by maternal udder-tilting. This frequently takes place as a maternal postural adjustment prior to milk let-down.

Ingestion

Given satisfactory stimulation, the lactiferous reflex, or milk let-down, occurs in the mother and facilitates the sucking reflex of the neonate. These two functions implement ingestion. Sucking is self-reinforcing and this assures proficient ingestive behaviour. The latter, in turn, establishes the supply of nutritive material for the ethological and physiological demands of the rapidly developing infant animal.

PLAY

Infant play appears to be important in the development of early social organizations, but solitary play is also important as a form of exercise. Play functions as a means of practising and perfecting adult behavioural skills necessary for defence and certain patterns of movement. Animal play is clearly valuable for the development of normal behaviour. It occurs most often in healthy young animals and its absence may be an indicator of reduced health. Social play, such as chasing and mock fighting, is most common. Solitary play can take the form of running around in confined areas. Elements of sexual behaviour appear in infantile play, of lambs for example, as early as two weeks of age. On these occasions lambs will mount each other, clasp and perform pelvic thrusts. Mouthing and biting at play feature commonly in foals. This may become aberrant biting if the infantile behaviour is encouraged to persist into adulthood (see Fig. 4).

9 Species Development

FOALS

Foals need about 30 to 50 minutes to attain a secure upright stance after birth and they usually fall about three or four times in standing attempts before they have success. The time taken to stand is variable, especially in the Thoroughbred, in which breed neonatal activity may be influenced by genetics and management. Some very viable foals can stand in 20 minutes while others, which could be considered abnormal in this respect, require two hours or more. Pony foals take an average time of 32 minutes to stand. Equine neonatal activity has been studied by the author, principally in Thoroughbreds, and the results, taken as norms of development, are shown in Tables 1 and 2.

Table 1. Normal sequence of equine neonatal activities in 315 foals

Activity	Order of occurrence	Time postpartum
Head lifting	1 or 2	1–5 minutes
Head shaking	1 or 2	5 minutes
Sternal recumbency	3 or 4	3–5 minutes
Ear erection and mobility	3 or 4	3–10 minutes
First rising attempt	5	10–20 minutes
First defaecation	6 or 7	30 minutes
Upright stance	6, 7 or 8	25–55 minutes
First vocalization	7 or 8	24–45 minutes
First sucking attempt	9	30–55 minutes
First successful suck	10	45–60 minutes
Second successful suck	11	1st suck +15–35 minutes
Third successful suck	12	2nd suck +5–15 minutes
First urination	13	90 minutes
First sleep	14	90–120 minutes

Table 2. *Quantitative features of equine neonatal activities*

Activity	Normal quantities
Falls in rising	3–6 times
Sensory inspections	1st: physical environment
	2nd: mare's anatomy
Number of sleeps	20–25 per day
Duration of sleeps	20–60 minutes
Number of sucks	18–24 per day
Duration of sucks	25–58 seconds

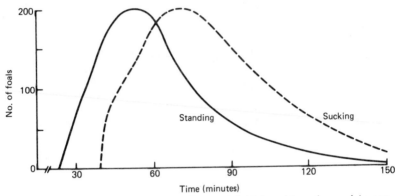

Fig. 24. Modal times for initial standing and initial sucking observed in 435 foals, mostly of Thoroughbred breeding.

Fig. 25. Early teat-seeking attempts in an Arab foal.

Fig. 26. Teat-seeking process in a foal, nearing completion about 50 minutes after birth.

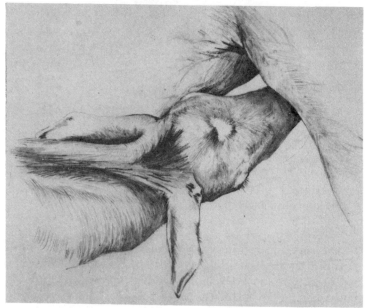

Fig. 27. Head posture in a day-old foal after completing all stages of teat-seeking.

The maturing foal gradually undertakes sorties of increasing distance from its mother and progressively spends more time playing with other foals as it grows. The equine bond, however, persists over one to three

Fig. 28. Posture of newborn foal defaecating for the first time; passing meconium required effort.

years and can remain strong even when later siblings are born. As the foal matures it sucks its mother less frequently, perhaps only about eight to ten times per day at six months. From one week onwards foals gradually begin to eat grass. They graze about 15 minutes per daylight hour by about three to four months of age. By one year of age foals spend about 45 minutes per daylight hour in grazing activity. At first, the young foal must spread and flex its fore legs to reach grass and, for a

Fig. 29. A day-old foal showing a perfect trotting gait.

week or two, until its neck has grown it is unable to walk and graze simultaneously. Foals spend most of the day resting during the first week. In the next two or three weeks they rest about half of the time. Foals typically rest in lateral recumbency until they are over six months, when physical and physiological maturity requires them to lie less often and to lie in sternal recumbency more frequently. Groups of foals may lie down together and social facilitation of resting behaviour is very evident.

Equine play is a good demonstration of play as a purely kinetic activity. For example, 75% of the kinetic activity of foals is in the form of

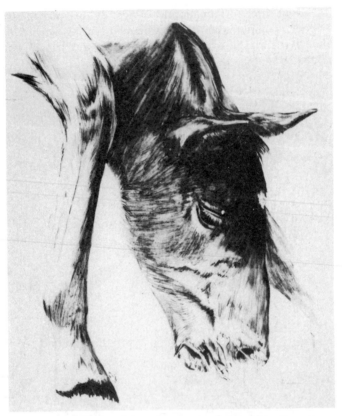

Fig. 30. A foal several days old, showing first grazing attempts.

play. Foal play begins by nibbling at the legs and mane of the mother. Mutual grooming occurs between the mare and foal at a later stage. The very young foal remains close to its mother, often maintaining physical contact with her even when walking. Week-old foals seldom go further than 5 m from their mothers. Among foal groups social play usually increases with age while solitary play declines. The latter is reduced to a very low order of activity by two months of age. Solitary play persists, however, in lone foals and their social play may relate to other animals and humans. Foals may also play with inanimate objects.

In addition to grooming with their dams, foals in groups also groom one another. Grooming bouts often initiate play and oral snapping

Fig. 31. Associating young Thoroughbred foals at play.

Fig. 32. Agonistic encounter between two associating foals, several months old.

actions are often seen in foals when they are initiating play. The commonest form of play between foals involves nipping of the head and mane, gripping of the crest, rearing up towards one another, chasing, mounting and side-by-side fighting. There are sex differences in foal play, with colts mounting more frequently and engaging in general play more vigorously than fillies. The response of fillies to colt play is often withdrawal or aggression.

CALVES

The newborn calf can stand within an hour and will usually commence nursing attempts within two hours. The nursing calf assumes a crouched stance with spread legs and shoulders lowered, allowing it to bunt upwards at the udder. This udder-bunting functions in the stimulation of milk let-down. The calf, when teat-seeking, nuzzles and licks along the under side of the cow. Higher aspects of the underline, such as the axilla and inguinal region, receive close attention but the calf will nuzzle any protruberance of the mother's ventral anatomy in the course of the udder-seeking.

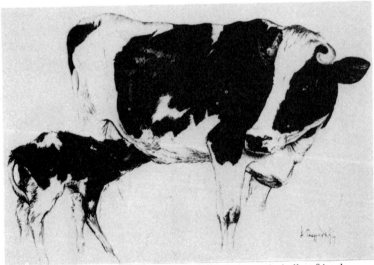

Fig. 33. Calf in first hour after birth exploring the underbelly of its dam as it commences teat-seeking.

Newborn calves normally nurse five to ten times a day, with each nursing session lasting up to ten minutes. The number of nursing bouts usually decreases with age, but this may vary depending on the rate of growth of the calf and the milk yield of the cow. Calves at six months of age nurse about three to six times per day. Dawn is a common nursing time. Other nursing bouts are centred around mid-morning, late afternoon and about midnight. Calves fed artificially with nippled milk hoppers show similar nursing habits. While sucking such feeders they often stand with their hindquarters deviated from the direction of their necks and prefer to touch a wall with their bodies in the course of sucking. This behaviour resembles the characteristic stance of the nursing calf with its mother in which it turns the side of its body and its hindquarters to be in contact with the side of the cow.

Mutual sucking is a common problem among bucket-fed calves raised together in crowded groups. Such inter-sucking can occur very frequently and cause skin irritation. prolonged sucking of the ears, the umbilical region or the prepuce is typical. Calves that indulge in inter-sucking are often unthrifty. Calves with wet ears from sucking by others may have them frozen in extremely cold weather. The persistence of inter-sucking behaviour in adult stock is not uncommon, especially in dairy herds. In a problem dairy herd as many as a third of the calves and a tenth of the cows may be affected. This behavioural vice is worsened by social facilitation by other milking cows. Since the condition is difficult to control, culling of offending animals is usually necessary and it therefore represents a serious economic problem. Some dairy farmers prevent the problem by penning young calves individually. Increasing attention is now being given to numbers of calves being fostered on nurse cows to feed naturally. When cross-fostering is attempted the normal procedure has been to present a cow, already in milk, with several young calves, perhaps newly born. Recent research on fostering has shown a much higher degree of success when the young calves to be fostered are presented to the nurse cow immediately following her parturition, before she has adopted her natural calf but while she is still in the 'critical period' of maternal awareness. At this time such cows readily adopt numbers of fostered calves and continue to facilitate their nursing subsequently so that these calves grow better. In selecting calves for fostering it is necessary to realize that calves which have not sucked naturally at all in the first six days of life are unable to nurse later on a lactating cow.

Calves play in a variety of ways, often cantering with their tails up in a typical fashion (Fig. 34). Playing calves also buck and kick out with both hind legs. Playful kicking, which is not aggressive, may be made to one side with both legs aimlessly, but at other times kicking at a given object, usually with one hind leg, may occur. Calf play also includes butting

Fig. 34. Tail-up cantering in a calf exhibiting play.

each other or inanimate objects, pawing the ground, goring bedding, threatening attendants and making snorting noises in the course of playful actions. Playful mounting behaviour is also commonly seen in calves. Bursts of play often occur when calves are released from confinement, when freshly bedded, when introduced to other calves and with other changes in their routine or environment. Play occurs most frequently in younger calves and in those in good health.

When calves are weaned they begin to show the clearly defined maintenance activities characteristic of adults. Maintenance activities, through lack of play opportunity, are modified in older calves penned individually. These calves kept in isolation may spend more time in standing rest than calves in groups, which appear to spend more resting time in recumbency. These adolescent cattle have the behaviour of their maintenance activities apportioned as follows: feeding 22%; drinking 2%; ruminating 28%; grooming 5%; resting 40%; and exploration 3%. Adolescent calves which have acquired behavioural schedules learn to anticipate feeding times and show restlessness as these times approach.

LAMBS

For the first few weeks of life lambs stay quite close to their own ewes. By one month of age young lambs spend two-thirds of their time in the company of other lambs. By this age play is well developed. The gambolling form of play in this species is very typical. The play involves upward leaps, little dances and group chasing. Play is reduced as lambs grow and is becoming rare by four months. In general it appears that lambs are born in an advanced stage of development, both physical and behavioural. For example they can stand, walk and nurse within the first hour. Their senses of sight and hearing are evidently better developed than those of some other neonates, such as the foal. The course of their behavioural development is therefore fairly rapid throughout the nursing phase.

PIGLETS

Within a few minutes of birth pigs can walk, see and hear. Certain physiological mechanisms, such as temperature regulation, are not mature at this time and temperature conservation by huddling is therefore a prominent feature of neonatal behaviour in the pig. Piglet mortality is high, commonly about 20%, and some of this mortality is due to misadventure such as wandering from the litter. Piglet wandering is an early sign of inanition. Normal healthy and well-fed piglets stay close to their littermates and the sow's mammary region. Wandering piglets are very liable to become crushed, chilled or traumatized by other pigs.

The formation of the social organization within the litter, which takes the form of the 'teat order', is a notable behavioural phenomenon which has received much study. The high degree of organization inherent in the teat order is an important means of litter survival. It also facilitates proficient synchronous nursing which is important in the prevention of piglet inanition. The teat order will be more fully described in the section on pig behaviour.

Piglets nurse frequently when they are young. Many retain oral prehension of the sow's nipple from one nursing bout to another. As piglets mature the number and duration of nursing bouts gradually decrease from about one every 20 to 30 minutes in the first day or two of life to about six per day at two months of age. Piglets attempt to eat solid food by seven to ten days of age. They are particularly attracted to solid food which is sweetened and formed into small pellets. Solid food intake, however, does not become substantial until about three weeks of age unless the piglets are milk-deprived. Young piglets quickly learn to eat the same solid food as their mothers and often attempt to share the sow's food with her.

Many piglets are raised under conditions of isolation in specific pig production enterprises. Their behaviour, in isolation from sows, is significantly affected. For example, those reared in artificial circumstances of isolation from their sows suck one another excessively. They defaecate frequently in the nesting area, in marked contrast to their normal discipline of excretion which develops rapidly over the first few days of life. Piglet excretory behaviour by four days of age shows clear preferential use of a communal and restricted excretory site. This is usually the corner of the premises furthest removed from the sleeping area. Isolated piglets give distress calls frequently when being handled, while conventionally raised pigs only give such calls when hurt. Even when separated from the sow for a few hours, piglets exhibit considerable distress by vocalization with either squeals or grunts. Vocalizations of distress increase with the length of isolation and, if isolation is imposed on piglets individually, distress vocalization is greater than if the entire litter is isolated from the sow as one group.

Early weaning of piglets at three to four weeks of age is often practised to increase the number of litters which a sow might have over a given period. These early-weaned pigs massage and nibble one another, particularly over their bellies. They spend less exploratory time rooting or nibbling inanimate objects. Early-weaned pigs in cages rest by sitting on their haunches much more frequently than do piglets on straw bedding and show a disinclination to rest in the normal lying position.

Piglet play develops notably in the second week of life. Play in young pigs is a very prominent feature of their total behaviour. It is mainly in the form of play fights (Fig. 35). Cheek-to-cheek fighting, in which each piglet bites and roots at the other's face, neck and shoulders, is standard practice. By several weeks of age chasing and gambolling are the

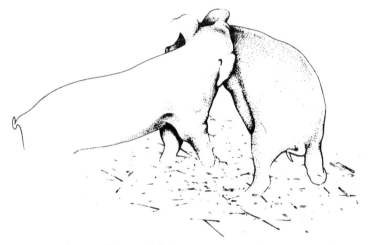

Fig. 35. Fighting in young piglets.

common forms of piglet play. The chases are usually brief. Playful individual behaviour involves rooting and mouthing of novel items. Oral manipulation and exploration of the environment is a major feature of pig behaviour, extending into adult life. Play in piglets is arrested during illness, as it is in other sick young animals. In this respect indices of play can provide parameters of health which could be put to greater use in the practice of preventive paediatric veterinary medicine.

The development of sleep has been studied in piglets during the first five weeks of life. The daily amount of time spent sleeping does not change over the five weeks and neither does the duration of sleeping episodes. Piglets sleep for about 26 minutes per hour and for the first five weeks of life this factor remains fairly constant. Cycles of sleep are not apparent in young piglets but on the average they sleep for about ten and a half hours per day. This contrasts with pigs of three to four months of age which sleep about eight hours per day. The duration of one type of sleep (paradoxical or REM sleep) does decrease significantly as piglets develop. Young piglets commonly sleep in a crouched position, with all four legs folded under the body.

CHICKS

The chick is active while still in its shell. Before hatching it attains an upright position with the head and neck elevated. It gives various calls within the shell; calls of distress and satisfaction have been identified. When the chick breaks through the shell it quickly seeks a source of heat, making characteristic calls as it does so. The natural heat source is the broody hen with which the chick makes very close physical contact after hatching. Thereafter close company with the hen is maintained as the chick matures and this provides opportunities for learning to refine the behaviour of maintenance. The bond with the mother is consolidated by imprinting, the sensitive period for which is between nine and twenty hours after hatching. The clutch of chicks with a hen maintains a close association with her and readily recognizes her physical characteristics and calls. When the process of physical development of the chick causes the down to be lost from the head, the hen rejects it. The clutch then becomes dispersed and more dependent on self-maintenance activity, while still associating with the flock in general. As they integrate with the flock, the chicks become established within its peck order. A stable flock of about 40 chickens maintains a stable peck order. Depending on the strain of bird, egg-laying commences by about six months of age.

TURKEY POULTS

The poult, like the chick, is active within the shell before hatching. After hatching the poult is very mobile. Imprinting appears to take place over a period of 24 to 48 hours. Imprinting errors with siblings, inanimate objects and attendants happen not infrequently. Great social cohesion within poult groups is notable from the first day after hatching but attachment to the hen turkey is also evident. Vocal and visual signals are used in the maintenance of close contact. As with the chick, this affinity facilitates learning certain critical activities, particularly feeding. Some artificially incubated poults are unable to initiate feeding or drinking and may die as a consequence of the lack of maternal association and the learning facility which this provides. Forced feeding can be practised with some success. Recent investigation of this problem has found that experimental poults, if exposed to lights flashing for about 20 minutes per hour for the first six days after hatching, fed more promptly and consumed more feed, particularly on the first day. A preference for green coloured lights was shown. It is interesting to speculate that the resemblance of this stimulus to the flash of metallic green from the colour of wild turkey plumage may be more than coincidental. After three months of age social hierarchies become formed in turkey groups and an established peck order is operational by five months of age, at which time groups show subdivision by sex.

Part III
The Behaviour of Maintenance

10 Introduction

Much of the behaviour of domesticated animals is concerned with self-maintenance. Highly successful self-maintenance is the basis of animal productivity. Activities involved in such maintenance appear to fall into eight primary generic systems. These are basically of innate origin and include much instinctive behaviour. Reproductive behaviour belongs in the additional category of gene or species maintenance.

Reactivity

A first line of defence in the individual is behaviour in reflex form. Reflexes require the integrity and full function of the central nervous system and rely on the peripheral nervous system with its specialized sensory transducers conveying sensory input. Reflex responses are typically prompt and brief, but exceptions occur. Their suddenness makes behaviours of this kind important to animal handlers, but their predictability is high, making preventive measures possible. Aggression is a main product of this system; vocalization is another product of reactivity.

Feeding

Alimentary behaviour includes eating and drinking as well as selectivity of food, grazing patterns and diurnally timed ingestion. Suckling, species-specific forms of food prehension and the behaviour of rumination can also be included in this system.

Exploration

Exploratory activities in general include investigative acts and empirical behaviour. The latter type of behaviour facilitates learning. Exploratory behaviour is most obvious in young animals but this whole category of behaviour is featured in all age groups, allowing learned behaviours to be acquired.

Kinesis

Certain patterns of action and movement apparently require regular expression. Some species, such as the horse, appear to have greater kinetic output than others, such as swine. The gaits of the walk, trot and canter are essentially similar in all the quadrupeds but some variations may occur. Various kinetic activities are tied in to other specific behaviours: perambulation in grazing, for example. Patterns of kinetic activity have been recognized in the ungulate fetus, close to term. Even under close constraint, characteristic actions of stretching and positional changes occur. Play activities are also included in this system of gross motor activity.

Association

Animals show behaviourally supported associations in a great variety of ways. Associations of dual bonds such as mother and young are readily recognized, but many other behavioural associations are featured in animal groups. Peck orders and other analogous hierarchial organizations of behaviour operate through social associations. Much social behaviour is dependent upon the behaviour of association and mediations of personal space. Association with man is a special form of animal socialization.

Body Care

This category is best recognized as grooming, but grooming is itself complex behaviour and fairly species-specific. Grooming occurs as licking, nibbling, rubbing, rolling, scratching, wallowing, sheltering, pecking, bed selection and huddling. These behaviours may be undertaken individually or mutually with others. The role of body care is probably considerable. This behaviour often ceases as a first sign of sickness at the clinical or subclinical level. The behaviour of evacuation is included in body care. In general body care deals with skin hygiene, thermoregulation and comfort-seeking behaviour.

Territorialism

Territorialism is a major determinant of behaviour and features in many activities. It is responsible for much agonistic behaviour, fight-or-flight responses, threat displays and herding behaviour. Territoriality is *'bourgeois'* in nature, as it entails proprietary behaviour in respect of defence of all or part of the home range of an animal. This defence is directed primarily against members of the same species.

Rest

Resting and sleep are forms of physical conservation which may occupy a quarter to a half of an animal's life. Since activity in animals tends to have fixed diurnal patterns, it is found that episodes of inactivity which permit rest and sleep have equivalent patterns. The existence of sleep in the domesticated animals has now become clearly documented, showing that both slow-wave sleep and paradoxical sleep regularly occur. These have been likened to sleep of the mind and sleep of the body respectively. Drowsing is another notable resting method. All forms of rest represent a major need in self-maintenance.

The several forms of behavioural systems essentially involved in self-maintenance are presented in greater detail in the chapters which follow. This systematic presentation of behaviour is used to create a basis for a general plan of behaviour, which allows the behaviour of the individual species to be similarly presented and examined.

11 *Reactivity*

REFLEXES

Many simple forms of behaviour occur as reflexes. Certain major reflexes involve extension or withdrawal of limb in response to various stimuli, locally applied. Limb reflexes have protective or postural functions in maintenance. Reflex evacuation, involving sudden defaecation or urination, is common in cattle and sheep following the stimulus of invaded 'personal' space. Orientation reflexes are seen in cattle, when their backs are turned to the stimulus of driving rain, for example. Seasonal and climatic behaviours are basically reactive in nature. Reflex escape struggles are reactions readily seen in animals placed suddenly in close restraint. Hyperactivity is a response often seen in animals subjected to a bombardment of stimuli, as, for example, in the course of shipment. Reactive vocalization occurs immediately following the separation of bonded pairs and in other forms of group disruptions. All these are evidence of the wide range of behaviours occurring as reactive responses, collectively comprising the system of reactivity in behavioural maintenance.

Agonistic behaviour embodies the behavioural activities of fight-and-flight and those of aggressive and passive behaviour, i.e. agonistic behaviour includes all forms of behaviour by an animal which is in conflict, physically or otherwise, with another animal.

Reactions to stimuli follow the basic organization of behaviour. By examining central nervous activity, a hierarchial system of controls can be observed. For example, if the bladder is full the stimulation of repletion passes from the bladder to the appropriate centres in the spinal cord. The responses are opening of the vesicular sphincter at the neck of the bladder, constriction of vesicular smooth muscle fibres in the bladder wall, adoption of the micturition posture and finally voiding of urine. These neural functions result from cord activity. This basic activity of the CNS can come under the control of higher centres via neural tracts, both afferent and efferent, in the white matter of the spinal cord which connects this basic control mechanism to a higher one

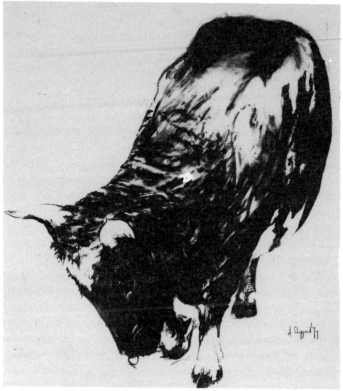

Fig. 36. The threat posture of a bull whose territory has been intruded. The posture is a manifestation of a temporary state of 'fight-or-flight'.

located in non-cortical parts of the brain. Higher parts of the CNS may superimpose themselves on the lower control centres, becoming operative when, for example, excretion is used for the purpose of territorial marking and for pheromonal dispersal. This would be a second step in the hierarchy of CNS control. A third step in the hierarchy would be shown by voluntary control of urination by the cerebral cortex. For example, many horses will voluntarily urinate in response to fresh bedding being placed in their stable or stall. Pigs also use cortical control over excretion.

Unconditioned reflexes, which are present as soon as the nervous pathways are developed, at or near the time of birth, are unlearned reflexes and are integrated in the spinal cord and brain stem. Conditioned reflexes, which are learned responses acquired by natural experience, are integrated at the level of the cerebral cortex. Much animal behaviour is complicated by conditioned reflexes. Some reflexes appear to be intermediate between the two, requiring the integrity both of the cerebral cortex and of the lower, brain stem integrating centres. For example, reflex following movements of the eyes appear to require the operation of both the cerebral cortex and the cerebellum.

The animal's willingness to move and the nature of its 'open awareness' are important aspects of reactivity. Various reflex responses may be general, such as the response to sound; specific, such as the response to a local stimulus on a given site on the body; or elaborate, such as responses to painful skin pinching, if circumstances indicate a need to establish safety. Some reflexes occur routinely in self-maintenance, such as in feeding and body care.

TEMPERAMENT

Reactivity is often dependent on temperament. The animal's perception and its awareness of its general environment are often related to temperament. The subject's appreciation of sensations such as vision, sound and position is qualitatively affected by its typical temperament. Movements of the eyelids and orbit are an important indicator of disposition. In the horse, for example, the upper eyelid showing contraction or elliptical change can indicate emotive temperament. Again, in any animal, a high mobility of the orbit is an indication of anxiety, while a very fixed orbital position may indicate distress.

In dealing with an unknown animal, therefore, observations of its attitude, disposition and temperament should be made before handling.

ESTROUS RESPONSES

During estrus, specific displays of receptivity are seen, in the mare and sow particularly. Sheep alone are lacking in overt estrous responses

Fig. 37. A mare being tested for estrus at a 'teasing board' with a stallion on the other side. In response to the stallion's stimulation (presence, vocalizations, bites) the mare has begun to show a normal behavioural display of estrus: the ears are turned back, the stance is stationary, the tail is elevated, the vulva is everted and one hind foot is slightly raised.

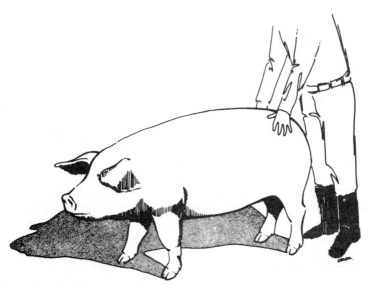

Fig. 38. The static posture adopted by a sow in estrus in response to firm 'haunch pressure'.

which can be induced by breeding techniques. While estrous reactivity is normally evident on minimal stimulation, there are increasing problems in determining this under intensive husbandry conditions. In such cases, estrous reactivity is improved by the composite stimuli from male 'biostimulations'. Increasing use is now made of androgenized female stock to function in a masculine teasing role to enhance estrous reactivity in larger populations of livestock intended for artificial breeding. The androgenized females themselves are remarkable examples of male sexual reactivity following therapy with testosterone.

AGGRESSIVE REACTIONS

Agonistic activities are principally involved in the determination of social dominance orders and relate both to the acts of positive aggression and to equivalent acts of submissive reaction. The aggressive acts, however, are most evident, particularly when the aggression is countered with aggression in an exchange between individuals closely matched in dominance status. It is common knowledge that mature male animals, such as bulls, boars and stallions, have a high level of aggressive reactivity in their behaviour. This is often displayed as prompt reaction to violation of personal space. Evidently testosterone is a chemical primer for reactive behaviour relating to instances of territorial threat. Threat displays are aggressive reactivity.

SUBMISSIVE RESPONSE

The reactivity of submission takes the form of characteristic gestures and postures. These may vary from the most common one of slight head depression with deviation away from the stimulus, to the gross display of catatonic submission in which the animal assumes recumbency and refuses to rise. This latter behaviour is a confusing condition in the presence of a concurrent illness contributing to recumbency (Fig. 40). The characteristic of such submission in cattle is vigorous, low extension of the head and neck when aversive stimulation is given. The recognition of submissive reaction is essential in handling sick or 'fallen' livestock of all species, to ensure that their circumstances are given appropriate consideration. General hypotonia is a common feature which, in its turn, is characterized by an abnormally low level of reactivity to such

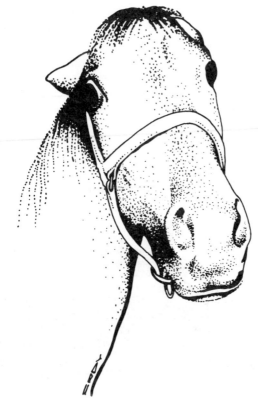

Fig. 39. Threat by a mare.

stimuli as are customarily effective in causing animals to change position or posture.

VOCALIZATION

The determination of reactivity involves keen and knowledgeable appraisal and is very much the special art of good stockmanship. In appreciating the responsiveness of animals, individually and collectively, to their husbandry the critical appraisal of their vocal sounds can be of considerable use.

The production of vocal sound is a major feature of reactivity. Many vocalizations are incorporated into responses concerned with alarm and threat. Associative responses also make use of phonation to varying degrees. Elaboration of vocal sound is a feature of communication in circumstances of social reactivity. As reactivity increases, when the animal is highly motivated, for example, vocalization tends to increase in volume, quantity and complexity. Many vocal signals are found as exchanges between mother and neonate, between breeding male and female, and by bonded individuals when separated.

Among farm animals, specific sounds relate to specific conditions of age, sex and circumstance. Hunger generates much vocalization, notably among young animals such as piglets, calves and chicks, but hunger in all age groups of farm animals causes hunger calls. Changes

Fig. 40. The state of submission in a 'downer cow'. This condition is analogous to cataplexy and may be termed 'tonic immobility'. Aversive stimulation will not relieve the condition, in most cases, and is more likely to exaggerate it.

in phonation, from major to minor key, take place in estrus in cattle and illustrate the effect of sex hormone on voice aids in communicating the chemical status of reproduction in the individual. Stallions are actively vocal in the breeding season; rams are particularly gutteral in autumn; bleating is almost continuous in goats during estrus.

Innumerable examples of specific vocalizations occur in the usual circumstances of animal farming, but there has been only a slow realization that phonation is a meter of reactivity. Vocalization in farm animals awaits more thorough recognition of its nature and importance.

SEASONAL BEHAVIOUR

Seasonal responsiveness to photoperiodism is sometimes evident in major changes in behavioural routines. In the seasonal breeding animals, such as horses, sheep and goats, the introduction of reproductive activities affects other routine behaviours, modifying some of their priorities. This feature of reactivity is more fully described in an earlier section and in reproductive behaviour, but since male animals, in their

Fig. 41. Concentration of breeding in a herd of free-living ponies showing synchronous response to the environmental, seasonal pressures of life on a Shetland island. (From a 19th century oil painting.)

breeding season, show increased aggression, the phenomenon warrants mention here. Seasonal changes occur in grazing behaviour. Sheep in winter show progressively more active grazing throughout the day, reaching the height of their diurnal activity prior to evening twilight. This grazing pattern does not contain the active phases of early morning grazing, characteristic of summer behaviour. In many parts of the world behavioural responses to weather are, indirectly or partially, associated with seasonal conditions, while being climatic reactions. Synchronized population or herd responses in feeding and breeding are examples of this.

12 *Ingestion*

Farm animals ingest vegetation using a well motivated system of regulated activity which represents their chief occupation. The on-going consumption of nutritious plant material and water is required to meet enormous and complex metabolic needs. These needs establish the ingestive drive. This drive directs the behaviour of eating and drinking. Herbivorous feeding behaviour is compounded by quantitative demands, diurnal rhythms, selectivity, fluid intake, digestive requirements, competition, oral mechanics, grazing techniques and daily schedules. The behavioural planning in all of this is, of necessity, fairly exact. Substantial provisional planning for the system of ingestion in the species is evidently set out in the genetic code. The continuous and specific neuroproduction, necessary for the design and support of all of this behaviour, is therefore innately programmed, giving species-specific feeding behaviour. By the superimposition of learning, there is ample opportunity for ingestive habits to be formed, modified and controlled through natural experience or through the dictates of husbandry. Adjustability to various feeding methods is not the least important feature of ingestive behaviour in farm animals.

INTAKE

The main features of the ingestive drive are more specifically referred to as hunger and thirst. The drive centres, governing the motivation and the regulation of these, are located in the hypothalamus. These are individual quantitative features, however, and when quality and complexity are taken adequately into consideration, in appetite for example, cortical contribution to a comprehensive ingestive drive organization is seen to exist.

A broad generalized neurological design of behavioural motivation and modulation is not unique to this system. It is seen in much behaviour which is clearly related to homeostasis, the maintenance of stable internal physiological states in the presence of a variable environment. In homeostatic goal-directed behaviour, specific body needs are met. These needs are mainly physicochemical. Ingestive homeostasis warrants some attention. Energy intake must meet energy expenditure for a steady state of body calorie-fuel balance. The laws of thermodynamics dictate that, in this state, the total body calorie-fuel

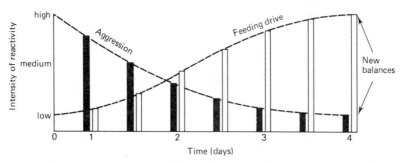

Fig. 42. Interference of instincts (potentiation by inhibition of antagonism), showing inversion of aggression and feeding intensity in newly formed groups of pigs (at time o).

input equals internal heat production (metabolism), plus external work (behaviour), plus maintenance of energy reserves (fat). If the balance is unfavourable there is weight loss. If it favours input of the energy-containing ingested food, the result is bodily gain such as growth or weight increase. Food intake is clearly the dominant factor which is subject to control automatically to maintain a stable or favourable balance of energy between ingesta and body. The neural structures primarily concerned in the control of food intake are several groups of neurons in the hypothalamus serving as centres. The lateral hypothalamus contains a feeding centre which stimulates the efferent output of ingestive behaviour. It controls the final motor acts of eating and associated behaviour such as food seeking and selection, typical of grazing, for example. The ventromedial hypothalamus contains a satiety centre. Eating is promoted unless the midline centres of satiety inhibit

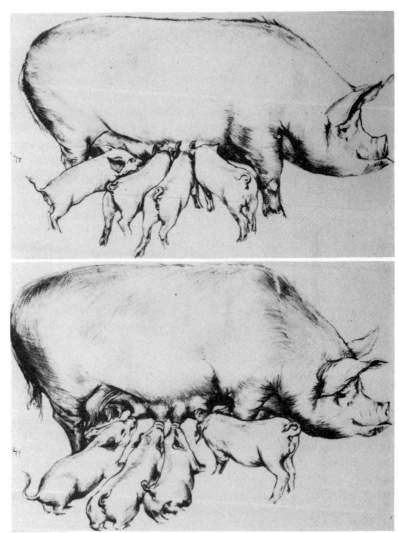

Fig. 43. Piglets sinking from standing to resting postures as they engorge with milk; the smallest piglet, on a posterior teat, sinks first.

the outer centres. These hypothalamic centres involved in control of food intake serve only as integrating units which process the incoming and outgoing stimulation. Clearly the incoming stimuli provide the critical information, firstly about the body's need for food and secondly about external food availability and its sensory stimulation. The present belief is that food intake is basically a maintenance procedure which is fairly continuous unless turned off by satiety signals. Animals evidently seek feed, not because they suddenly experience hunger, but because satiety signals have stopped being generated. In the determination of satiation, gastrointestinal signals do not constitute the major factor, though they may play a modifying role. Nor does a repletion-detecting system have a major influence on suppression of feeding behaviour. The present concept is that special receptors in the brain detect the levels of blood-borne clues of nutrition such as glucose, lipids, a satiety hormone and metabolic heat. Evidence of the existence of such a feed-back of significant information has been shown in experimental animals. The homeostatic system of ingestion can be seen to exist as a functional and potentially autonomous arrangement through all these mechanisms.

Many paradoxical circumstances are encountered in the feeding behaviour of farm animals, however. For example, a single pig will not eat as much as when it is one of a pair. A pair of pigs will, in turn, eat somewhat less than if they are part of a larger group. Feeding motivation, therefore, can overrule homeostatic principles. Highly successful feeding motivation, or feeding drive hypertrophy, is a major key to farm animal productivity. Physiological facts alone do not explain all about feeding behaviour. Feeding behaviour is strongly influenced by reinforcement, both positive and negative, from food palatability and by the environmental and social associations of feeding. It is necessary, then, for the concepts of motivation, drive and reinforcement to be incorporated into any comprehensive view of food intake control. Inheritance of behaviour too must be considered. It has been observed, for example, that animals such as some piglets and calves which are observed to have good appetites and to be very active feeders when young, usually become good growers and adult producers. They are usually the progeny of families of stock of similar constitution.

SPECIFIC APPETITES

A constant question, relevant to ingestive behaviour, is whether homeostatic mechanisms exist to stimulate consumption of specific and essential nutrients, in proportion to their need by the body. The answer appears to be in two parts. Firstly, regulatory systems exist for water and sodium, creating thirst and salt appetite, both of which are very real.

Secondly, nutritional deficiences in general do not have homeostatic methods of self-correction although some can affect behaviour through creation of depraved appetite. The ability of the animal to correct a specific mineral deficiency, even when given free access to the necessary mineral, is poor. Essential organic substances also lack regulatory ingestive systems.

THIRST AND SALT APPETITE

Thirst is an occasional property of the ingestive drive. The brain centres mediating thirst are located in the hypothalamus. These centres can have a great quantitative influence on drinking, but the government of thirst is not totally vested in this arrangement. The regulation of water and salt is linked together in urine production and is under the influence of the antidiuretic hormone (ADH) of the posterior pituitary gland and aldosterone of the adrenal cortex. In the production of urine, water deficits or gains are compensated for by partially dissociating water excretion from that of salt, through changes in ADH secretion. Under the influence of ADH, kidney ducts have their permeability changed so that urine is concentrated and body water is conserved. Aldosterone regulates sodium absorption in urine production and thereby limits salt loss from the body. A hormonal substance, angiotensin, is produced by kidney cells to promote aldosterone production. The latter hormone plays a dominant role in maintaining the volume of extracellular fluid in the body. This complex balancing system is basic to the body's method of maintaining a critical level of fluid in all circumstances.

Angiotensin has another role in fluid regulation, since it can directly stimulate thirst. This is undoubtedly a major method by which drinking behaviour is stimulated when extracellular body fluid volume is diminished. Dryness of the mouth and throat is still another means to direct and control the thirst drive. This dryness evidently creates thirst and appears to be capable of metering water intake so that adequacy is signalled, even before the ingested water has had time to be absorbed from the gastrointestinal tract. In the horse, for example, the caecum is the main site of net water absorption, with the colon a secondary site for water uptake.

Salt appetite is another important homeostatic system which affects ingestive behaviour. Salt appetite is held to be innate in mammals and is described as having two components: a predilection and the specific appetite regulated by need. There is little doubt that farm animals like the taste of salt and will consume it, typically by licking, beyond their need. Sometimes, in pigs for example, voluntary salt ingestion is excessive enough to create a state of poisoning. In addition, farm animals have their drive to ingest salt very markedly increased as a result of a deficiency. For this reason, freely available salt licks for animals are put to full use. They provide one way of supplying trace elements which might not be ingested, even if made equally freely available in another mixture. Sometimes a salt appetite in farm animals can be very acute and can lead groups into long searches for salt. Grazing animals with access to the seashore can be seen foraging on the shore below the high tide line, where they will frequently ingest seaweeds and will lick and chew other salted material. This is evidence of the pleasure quality in salt appetite.

Excessive water intake has been observed in confined animals such as installed horses. The syndrome has been termed polydipsia nervosa and will be mentioned under anomalous behaviour.

GRAZING BEHAVIOUR

Diurnal patterns of eating are characteristic of grazing behaviour in horses, cattle and sheep. The distribution patterns of grazing periods are correlated with patterns of daylight. The actual duration of active eating is influenced by food quality and availability. Grazing activity is largely confined to the daytime and the onset of active grazing is closely correlated with the time of sunrise. Most of the daylight hours are occupied with grazing periods. These periods usually add up to more than half of total daylight time, but some night grazing is also practised. The most active grazing season coincides with spring in most grazing regions. The ratio of day to night grazing is affected by very hot weather in summer when more night grazing occurs. Cold and wet spells of weather in winter can reduce grazing, but they do not have a very significant effect on the ratio of day to night grazing. In winter, horses spend most of their time grazing while less time is spent on grazing during very warm weather. Summer grazing behaviour in both cattle and horses is adversely affected by heat and fly attacks. Both of these circumstances demand a behavioural switch to body care activities which preclude grazing.

On arid ranges sheep and horses have been observed to travel long distances each day to water. It is likely that usable range area is determined by the furthest distance from available water that livestock are able to travel on a daily basis. Grazing animals can ingest snow as an alternative to water if the latter is difficult to reach, or is frozen, in winter. Range grazing animals, in the presence of snow, can afford to forage outside the usual watered territory. Cattle drink twice daily, on

Fig. 44. Mare grazing while supervising her sleeping foal.

average, in warm weather, but once-daily drinking is more common in winter.

Grazing involves travel in addition to time. The nature of the grazing territory, and its quality, influence grazing travel. Horses may travel 3 to 10 km (2 to 6 miles) per day grazing and spend about two to three hours in grazing travel. Cattle move from 2 to 8 km (1 to 5 miles) daily in grazing travel distance and about two hours in grazing travel time. Sheep on range travel about 6 km (4 miles) per day and spend about two hours on this travel. On good pasture land, sheep may only travel about 1 km per day. Range livestock also travel considerable distances regularly to salt lick locations which should, therefore, be strategically and adequately sited.

The grazing activities of milking cows are arranged around the milking times. Very active grazing usually follows each milking session. Active grazing bouts are usually followed by rumination in sternal recumbency.

HOUSED FEEDING BEHAVIOUR

Most swine are permitted to feed *ad libitum*. Their daily feeding time totals about one hour. They may be fed controlled rations twice per day and under these conditions they show hunger if the feeding schedule is even slightly delayed. The permanently stalled horse spends about three hours eating and a total time of 15 minutes drinking about 36 litres (8 gallons) each day. It may spend up to half an hour licking at a salt lick. Cattle in loose-housing spend about five hours per day eating. Their rumination time is also reduced. Although cattle in feedlots are in a very unnatural environment they still show diurnal rhythms similar to those evident in natural grazing, though their total eating time is very reduced. In place of natural grazing bouts, feedlot cattle have about ten to fourteen feeding periods, with approximately 75% of these occurring during daylight hours. If hay or silage is fed, five hours per day may be spent on active eating, as in the loose-housing system. Eating time becomes reduced as roughage is reduced and the proportion of concentrate feed is increased. Carefully managed dense groupings of feeding animals have the propensity for creating hypertrophy of their feeding drives, causing increased ingestion and resultant productivity.

SPECIAL INGESTIVE FEATURES AND PROBLEMS

In the course of their development, young mammals proceed from ingestion by milk sucking to ingestion by chewing. Young foals grazing with their mares will begin to eat grass tops as their capacity for grazing develops. Nursing foals and piglets are quick to develop the ability and drive to ingest dry feed presented as a supplement to natural milk. This infant feed is usually an expensive compound and when it is made available to them it is usually behind a barrier, below which they can creep, but which prevents the mother's access.

Goats can graze as sheep and cattle do but they have a strong drive to browse. Browsing is the ingestion of bush and tree vegetation. Given free range in bush land, goats will spend about equal times browsing and grazing. Horses also browse occasionally and one form of this is eating bark, very small trees, branches and leaves. Bark eating is likely to be adopted by each horse in an enclosed group with tree access. This leads to trees being debarked and 'ringed' from root level up to a height of the highest reach of the horses. The propensity for wood-eating in horses could be considered as vestigial browsing behaviour. In certain forest regions equine browsing could serve beneficially as auxiliary or supplementary ingestive behaviour. It is clear that most farm animals indulge in some occasional 'supplemental' feeding. It is uncertain, however, if this represents empirical searching for rare nutritional elements, but it is thought to be sometimes responsible for the ingestion of foreign bodies, such as those which create 'hardware disease' in cattle.

Depraved appetite, or pica, is a notable feature of phosphorus deficiency in cattle. It features in the chewing of wood, bones, soil, etc. At first glance it would appear to be a homeostatic ingestive behaviour, but this is misleading. Even when given free access to bonemeal, deficient animals still do not convert the pica to selective ingestion of the appropriate foodstuff. When phosphorus-deficient cattle eat such bonemeal they seldom eat enough to correct completely a deficiency great enough to have caused the pica. Horses have been found incapable of correcting mineral deficiency when given free access to a digestible mixture rich in the necessary mineral.

Although horses might appear to eat fastidiously they nevertheless develop more eating problems than other farm livestock. Wood-eating has been mentioned, but more depraved appetites may develop as habits. Examples include the eating of bedding, dirt, sand, gravel, tail hairs and faeces. In addition, some horses develop the habits of swallowing air, bolting their feed and of drinking water excessively. Most of these ingestive vices are potentially dangerous to the horse since they can cause various types of colic. Some of these ingestive aberrations will be given further discussion in the sections on equine behaviour and anomalous behaviour.

13 Exploration

Animals show strong motivation to investigate environmental contents. This motivation is normally lost, however, when the environmental feature has become familiar. Yet the basic exploratory drive appears to maintain a potential for generating activities which focus the senses of the animal upon additions, changes, salient features and novelties in its close environment. Ready loss of investigative value in neutral stimuli allows the drive to be conserved. The reserves of explorative behaviour equip the animal with a system of behavioural adjustability which can be brought readily into operation.

The sensory pick-up, which follows exploration, usually facilitates the output of pre-set behaviour which can be selected from the animal's stocks of behavioural response patterns. The efficiency of the exploration is vital to the production of an appropriate response. Continual behavioural response adjustability within its environment is obviously of major importance to the free-ranging animal. It can be considered to be a very elemental feature in its self-maintaining behaviour. It is evident also that such activities are supported by investigative and exploratory behaviour in general. In systems of confined husbandry, exploratory acts are reduced and the innate tendency to explore is likely to lack outlet in an environment of minimal variety.

Drives emerge from needs and the exploratory drive apparently stems largely, though probably not exclusively, from sensory perceptive needs in the animal. Perceptive needs are met with variation in environmental stimuli. This is the first feature commonly lost to the animal in restrictive confinement systems of husbandry. Under circumstances of very close and chronic confinement the exploratory drive, in due course, often becomes redirected and produces alternative behaviour which is likely to be a simple but maladaptive activity. Drive reduction occurs with need reduction. When the perceptive need is not met, persistence of the exploratory drive, or its redirected analogue, usually occurs. This creates the repetitions characteristic of stereotyped behaviour.

The role of investigative behaviour in the facilitation of learning cannot be ignored. By exposing the animal, in a state of heightened awareness induced by the original stimulus, to the new circumstances, the animal's mediation with such circumstances establishes experience. Experience is a valuable commodity in animal behaviour. In many cases the experience will have involved the use of empirical behaviour in the animal's endeavour to achieve resolution of a given episode. Empirical activities therefore are also incorporated in this class of behaviour. Some empirical activities involve interactions with the inanimate environment using species-specific characteristics: horning of trees, bushes and turf; scraping earth and snow; nosing artifacts closely to smell and to touch with the upper lip; rooting into soil and bedding; head-pressing on fenced perimeters for movability; leaning on upright structures for yield; licking hard surfaces; and overseeing from prominences.

Empiricism in behaviour is trial-and-error activity. Empirical activities are of particular value in the practice of behavioural features which develop as the animal grows and matures. It is therefore a key ingredient in ontogeny. Continuity in a given series of trial-and-error actions usually brings improvement in the performance of patterns of behaviour. Examples of this can be most clearly seen in the behaviour of the long-legged, short-sighted newborn foal. The foal's rising attempts, which begin soon after birth, involve coordination, balance, effort and change. Several failures to stand up usually occur, with the foal falling back down; but with successive trials improvements occur. Soon the foal is able to rise to a fixed, partly upright, posture. Thereafter the ability to rise and stand is achieved. The trial-and-error process of rising from a lying to a standing position is completed (Fig. 45).

Although this progressive behaviour illustrates learning, it also illustrates that empirical activity can be innate. Again, when the foal is ambulatory, its exploratory activity becomes engaged in 'teat-seeking' (Fig. 46). By trial-and-error activities the foal locates and investigates the mare's limbs and ventral regions, the mare's inguinal region and ultimately her mammary gland. The trial-and-error activities in the foal's teat-seeking behaviour are often extended over a substantial period of time. With highly successful foals it may be complete within thirty minutes. In others it may take several hours. In still other, rare instances, there may be total failure. When the trial and error activities

Fig. 45. Trial-and-error attempts at rising in a newborn foal.

Fig. 46. Trial-and-error teat-seeking in a newborn foal.

terminate in success, they do not require to be reproduced on the next teat-seeking occasion.

Although the neonatal foal is a good model for empirical or trial-and-error behaviour, neonatal behaviour in general is evidently rich in empirical activity. Within herds, young animals learn the identities of their own mothers with the aid of trial-and-error exploration. The teat order in young piglets is quickly mastered through trial and error. Social relationships and hierarchies, in all animal groups, have been determined, essentially, through empirical activity. The support for such empirical behaviour is basically the exploratory drive for, while this stems principally from perceptive needs it is also generated by other, even less tangible, needs such as self and status determination.

Animals show both inquisitive and acquisitive reactions to recently familiarized exploratory findings. Returning to young foals for an example, it is seen that they are normally inquisitive towards people but acquisitive in respect of their mares. Lambs are inquisitive towards other lambs but acquisitive towards their own ewes. Young piglets are inquisitive towards small environmental items but acquisitive towards their litter group as a whole. Pasture breeding sires, such as bulls and stallions, are inquisitive and acquisitive towards the same individual breeding female in pro-estrus and in estrus respectively, determining the particular state by exploratory sexual behaviour. Newly calved nurse, or suckler, cows are highly inquisitive towards any young calf, even a calf-call, but very acquisitive towards their own calves only. Biology recognizes variations; some variations make exceptions, and some exceptions to these general behavioural characteristics naturally occur.

A general exploratory system is evident throughout many animal activities. This system can be outlined most simply as an organization of motivating cycles and consequent activities as follows.

1. Needs of sensory perception and critical determination, heightened by incomplete but key stimulation, activating the related drive.
2. The exploratory drive directing engagements in exploration, investigation and empirical interaction with the environment.
3. The receipt of mass sensory feed-back to meet the original needs.
4. Drive reduction obtained from the appropriate behaviour produced in resolving the event.
5. Return of the system to a basal level of function with lodgement of the circumstantial events in short- or long-term memory storage.

Fig. 47. Trial-and-error exploration in a group of female goats following
separation from their kids for a one-hour period.

Fig. 48. Exploration of kids for identifiable characteristics following a period of
experimental separation.

More comment on the perceptive need in animals is warranted. Many needs are elemental, such as food, water and air. Some are more complex, such as territory and kinetics. Other needs are compound, such as association and body care. The perceptive need is more abstract. It is, to some extent, polyvalent. The perceptive need is not a singular one; it is probably a holistic need, where the requirement for the multi-sensory system is an absorption of quantity and quality of tolerable ambient properties. Deficits in some aspects may be largely compensated by the balance of the remaining multi-sensory perceptual system, particularly when this is subject to variability. The senses of sight, hearing, smell, taste and touch feed the perceptive need of the animal, but even all of them together may not meet this need, in a chronic situation, if the sensory composition is inadequate. In an attempt to summarize, it may be suggested that it is not only a sum of senses, but a sensory composition, which is the perceptive need in higher animals.

The social animals evidently have to acquire their own terms of reference. They evidently need to determine their own characteristics, abilities, limitations, social status, etc., since they expend much activity investigating these factors. They do this through social explorations and constant empirical interactions with their animate and inanimate environments. Animals constantly test their powers of dominance with others in encounters of petty aggression. The use of force tests their powers of reach and prehension in feeding. They can determine their social tolerance in mutual grooming. They can learn of their strength through pushing. They can also learn, through play, their personal powers of flight, attack and defence. The quite basic need of self-determination, or identification, is one which is closely allied to the perceptive need. Inquisitive needs use the common exploratory drive to seek their goals.

Empirical social interactions are evidently used in relative self-determination and, probably for this reason, livestock held in social isolation do not usually develop in the same way as others of their kind. It has been said, for example, that a horse confined on its own is in bad company. Solitary animals can all be considered to be in poor circumstances to have needs of determination met. Acts of anomalous behaviour, such as pathological 'mouthing', occur relatively more often in animals confined alone than in others. These behavioural disorders might be other examples of anomalies resulting from a redirected exploratory drive of an unmet inquisitive need.

In addition to anomalous behaviour, the exploratory drive is involved in frank clinical behaviour. Aberrant exploratory activities are common in rabies. Exploratory behaviour becomes significantly suppressed in many illnesses in animals, particularly those which have depression as a component of a syndrome. Indeed, the opinion could be held that absence of all exploratory behaviour is diagnostic of clinical states of depression in animal disease.

14 *Kinetic Behaviour*

Some behavioural activities seem almost purely kinetic. The mechanics of motion, for example, are chiefly used to transport the animal. Movement is vital in pastoral living and the farm animals have a legacy of substantial locomotor demands which they have met through their hooves.

GENERAL KINESIS

Many kinetic behaviours have patterns, rhythms and phases in their expression. Phases of activity are seen at dawn. Kinetic patterns are seen in reproductive behaviour and in modes of play. The rhythm of gait is classical. Purely kinetic behaviour is best seen in the horse but it exists in lesser degrees in cattle and sheep. The system of kinetic behaviour has range, rhythm and pattern. This system is of innate origin, being a feature of the species. A basic requirement of the system is territorial space, akin to the natural habitat. The innate kinetic drive is modified by diverse factors. Acting in apparent conflict are habitat loyalty, or home range bond, and the periodic inclination to roam; but these apparent conflicts are both forms of ecological mediation.

The need for kinetic competence in pastoral animals is considerable, since they must graze proficiently under a variety of conditions of location and weather. The daily water requirement adds a further kinetic role. Horses at range, for example, may travel up to 65 to 80 km (40 to 50 miles) daily to water. Sheep will also travel great distances regularly to water where vegetation and water are both scarce. While these travelling demands do not usually exist in conventional farming, nevertheless this travelling ability is shared, to some degree, by all grazing livestock.

In cold or wet weather livestock conserve their movements, although horses are sometimes very active in cold weather, frequently galloping about their territories. It is believed that this is done to generate metabolic heat. Horses, cattle and sheep periodically attempt to travel or roam away from their home range. This phenomenon is quite well recognized but has been poorly documented. Given the opportunity, they will sometimes travel at the walk for considerable distances. One chore in herding sheep, for example, is to return them periodically to their home grazings after roaming. Horses and cattle also roam, even leaving adequate grazing conditions to do so. One feature of note in this travelling behaviour is the 'Indian file' order of following a lead animal. The social identities, or other constant characteristics, of these lead animals have so far been difficult to determine. Animal routes are often seen, from the air particularly, as pathways which criss-cross grazing ranges. They are even found in most paddocks. Animal walks are not, therefore, reserved for dramatic excursions but occur frequently, even regularly, in normal domestic conditions.

Ambulatory kinesis is, of course, incorporated in grazing. To this extent, the grazing rhythms described are also kinetic rhythms. Kinesis is also a prominent component of play behaviour. Developmental phases contain many specific patterns of kinesis and these have been described within the context of ontogenic behaviour. Such patterns of kinesis are evident in simple fetal movements, complex fetal movements, fetal righting movements, neonatal exploration, infant play, juvenile interactions and social exchanges. Though many of these kinetic phenomena are pronounced activities, the principal kinetic output is routine locomotion.

LOCOMOTION AND GAITS

In locomotion the limbs act synchronously in any one of a variety of patterns, each of which is termed a gait (Fig. 49). Two forms of gait pattern exist, symmetrical and asymmetrical. In symmetrical gaits the movements of limbs on one side repeat those of the other side, but half a stride later. In asymmetrical gaits the limbs from one side do not repeat those of the other. Symmetrical gaits include the walk, the pace and the trot. Asymmetrical gaits include the various forms of the canter and gallop, including the lope and the rotary gallop.

The full cycle of movement of a leg during the support, propulsion and flight phases is termed a stride. A stride is a full cycle of movement of all limbs, while stride length is the distance covered between successive imprints of the same hoof. The sound produced when a hoof strikes the ground is the beat. If each limb strikes the ground separately the gait will be a four-beat gait. If diagonal limb pairs are placed down simultaneously, as in the trot, the gait is two-beat since only two beats

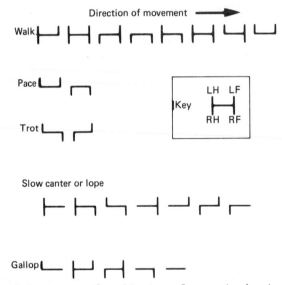

Fig. 49. Typical sequences of combinations of supporting legs in mammalian gaits.

will be heard for each stride. The canter is a three-beat gait—in waltz time. A lead leg is that leg of either the fore or hind limb pairs which leaves the ground last during the canter or gallop. The gallop, of course, resembles the canter, but since it is faster it has an extra 'floating phase'.

The gaits of the walk, the trot and the canter are the forms of quadruped motion fundamentally involved in all locomotor behaviour.

Walk. The walk is defined as a slow regular symmetrical gait in which the left legs perform the same movements as the right, but half a stride later and in which either two, three, or sometimes four legs support the animal at any one time. The support role is more important in the fore legs, which are nearer the centre of gravity, and the propulsive role is more important in the hind limb. Approximately 60% of the static weight is supported by the fore legs in the horse. Within each stride each limb for a time acts in a support phase and in a non-support or swing phase. In the walk the support phase is longer in duration than the swing phase and determines the stability of this gait. As the walk speed is increased the duration of the support phase decreases while that of the swing phase increases.

Trot. The trot is a symmetrical gait of medium speed in which the animal is supported by alternating diagonal pairs of limbs. The fore limbs are free of the ground longer than the hind limbs to allow the front feet to clear the ground in advance of the placement of the hind feet, on the same side. If there is a period of suspension between the support phases the gait is referred to as a flying trot. A slow, easy, relaxed trot is called a jog or dog trot. Sometimes the term dog trot is used to mean that the animal travels in a straight line with the hindquarters shifted to the left or right. The trot is occasionally classified also as ordinary trot, extended trot and collected trot. Standard bred horses use the extended trot in racing when the limbs reach out to increase stride length and speed. The hackney uses the collected trot which is characterized by flexion and high carriage of the knees and hocks.

Canter and Gallop. The canter is essentially a slow gallop. It is a three-beat gait with one diagonal pair of limbs hitting the ground simultaneously. The hoof falls are typically as follows: one hind foot, the other hind foot and the fore foot diagonal to it simultaneously, the remaining fore foot. The canter is a gait in which the horse can use its neck muscles to advantage by the accentuated upward swing of the head which helps to raise the fore quarters and to advance the fore limb. As the cantering horse tires it will bob its head more to utilize this minor auxiliary system.

The gallop is an asymmetrical gait of high speed in which the animal is supported by one or more limbs or is in suspension during parts of the stride (Fig. 50). If the limbs are placed down in a circular order, e.g. RH–LH–LF–RF, the gait is referred to as a rotary gallop. If the hind limb leads through to a fore limb of the opposite side, e.g. RH–LH–RF–LF, the gait is termed a transverse or diagonal gallop. Horses seem to prefer the traverse gallop. In the transverse gallop the placement of the limbs, and therefore the support pattern, is transferred from the lead hind limb diagonally to the fore limb of the opposite side. As in the

Fig. 50. Racing horses showing two critical phases of the gallop. The near horse has the off-fore leg in the 'support phase'. The far horse is in the phase of 'suspension', with all limbs in the 'swing phase'.

rotary gallop two forms of this gallop exist, dependent on the order of foot falls, e.g. LH–RH–LF–RF, a right lead transverse gallop, or RH–LH–RF–LF, a left lead transverse gallop.

KINETIC REQUIREMENT

Respiratory action is tied to running. The oxygen consumption of a running animal increases linearly with the running speed. The net cost of running is constant for each species but decreases with increasing body size. The fact that a large animal can move one unit of body weight over one unit of distance more cheaply than the small animal must be considered of some advantage for large body size. The walking horse is an example of a most economical form of motion, since the metabolic cost of moving, as opposed to not moving, which is the net cost of locomotion, is remarkably small in this animal.

Other kinetic actions include various forms of stretching (Fig. 51).

These are usually performed after rising. Most stretching occurs as a series of actions as follows: flexion at the throat, arching of the neck, straightening of the back, elevation and movement of the tail and full extension of one or other hind limb. Extension of the fore limbs, singly

Fig. 51. Upward stretching kinetics in a young horse.

or together, is a related exercise. All forms of exercise serve as substrata to the principal forms of kinesis. In its turn, kinesis serves as a substratum for the majority of behaviours. Voluntary kinesis indicates that it has some of its own motivation, evident in a general kinetic drive. This drive may be more basic than some others, yet it tends to be often overlooked in considerations of the ethological needs of animals. Even

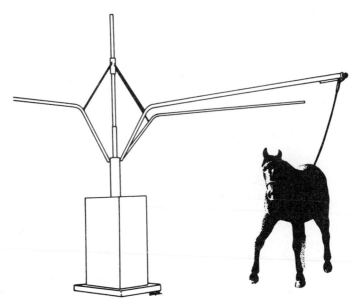

Fig. 52. Mechanical exerciser for horses facilitating kinetic output.

the productive farm animals require opportunities for exercise. Horses certainly need exercise on a daily basis. It is now clear that severe kinetic restrictions result in gross forms of anomalous behaviour.

15 *Behaviours of Association*

Numerous behaviour patterns occur in domesticated animals in the course of social interactions. Pair bonds are notable associations. Social relationships between and within sexes and age groups are also prominent in each species. Among enclosed animals, modified social behaviours are seen. The domesticated animal's social interaction with people varies considerably from species to species, dependent on the system of husbandry and whether the animal has received a traumatic human association. This form of special relationship involving the positive association of the animal with man is termed socialization.

SOCIAL BEHAVIOUR OF FARM ANIMALS

Farm livestock associate together in groups. Even under free-range farming systems voluntary grouping is very evident. Sheep, cattle and horses maintain visual contact. Swine show more body contact and keep in auditory communication. When disturbed suddenly, sheep and horses first bunch together and then run in a group from the source of disturbance. Pigs and cattle move in looser groups. During the bunching of animal groups in natural or high-density situations, individuals may be forced to violate the personal space of others. Social interactions at such close quarters depend on the position of the animals in the dominance order. Dominant and subordinate postures and responses are appropriately adopted. This stability of association requires recognition between individual animals, established social positions and memory of social encounters which establish social status. In groups of pigs, cattle and horses aggressive behaviour is most seen when these are first formed. Frequent changing of group members should be avoided in farming practices. Production of milk and other physiological functions can be affected for several days while aggressive social interactions are taking place in a radically changed group. Although sheep seldom show overt social dominance, they may show aggressive butting if intensive husbandry conditions increase competition over food or bedding areas. Butting in cattle and sheep, biting of mane or withers in the horse and pushing, biting and 'rooting' in pigs are the common forms of initial antagonistic associations. The

Fig. 53. A social exchange in a pair of heifers in a herd of dairy cattle.

introduction of individuals to established groups is hazardous due to the mass of agonistic encounters the alien receives. An estimation of the total number of group members that can be recognized or remembered by each individual appears to be 50 to 70 in cattle and 20 to 30 in pigs.

Social Dominance

Piglets show some competitive fighting within a few hours of birth for the preferred anterior nipples of the sow. Other farm species do not develop a stable social order until some time after weaning. In semi-wild cattle, bull calves stay in cow herds, dominating the females up to about two years of age and then move into 'bachelor' bull groups. Social dominance effects can be very important in cases of high stocking densities or poor farm design. Inadequate trough space, narrow races, inadequate space in indoor housing or lack of feeders can mean that dominant animals command resources at the expense of subordinate

Table 3. *Outcome of agonistic encounters in a group of five beef cattle showing the development of a linear social dominance order*

Pair comparison	Total encounters	Wins (%)	No. of subordinates	Social dominance order
263 vs. 39	63	96·8		263
263 vs. 1157	27	81·5		↓
263 vs. 7102	39	97·4	4	39
263 vs. 925	69	95·6		↓
39 vs. 1157	23	87·0		1157
39 vs. 7102	37	97·3	3	↓
39 vs. 925	29	82·8		7102
1157 vs. 7102	13	76·9		↓
1157 vs. 925	12	66·7	2	925
7102 vs. 925	26	61·5	1	

Table 4. *Outcome of agonistic encounters in a group of five beef cattle showing the development of a social dominance order with a triangular component*

Pair comparison	Total encounters	Wins (%)	No. of subordinates	Social dominance order
135 vs. 3362	46	97·8		135
135 vs. 2568	55	89·1		↓
135 vs. 2067	69	97·1	4	3362 ←
135 vs. 2772	30	100·0		↓
3362 vs. 2568	30	93·3		2568
3362 vs. 2067	42	95·2	2	↓
3362 vs. 2772	20	35·0		2067
2568 vs. 2067	67	76·1		↓
2568 vs. 2772	26	80·8	2	2772 —
2067 vs. 2772	14	78·6	1	

animals. The latter will suffer and health and general production can be affected. Documented examples include the higher internal parasite load carried in some subordinate goats and higher death rates in subordinate stock during droughts when scarce food was commandeered by dominant stock.

Since the horse is very responsive to small changes in stance or skin

pressures, these cues, used during dominance–subordination interactions, are also utilized by horsemen in controlling their animals.
Sometimes tranquillizers have been used to aid social tolerance when
strange pigs have to be penned together.

Leader–Follower Relationships

Pigs are reluctant to lead and require to be driven, but cattle and horses
are all subjected to a leader–follower order in free-range conditions.
In naturally constituted flocks of sheep constant leadership is not
common but the oldest ewe may tend to lead. In groups of dairy cows
the mid-order animals in the social hierarchy usually lead. Some use is
made by man of the 'Judas' animal to lead groups in to slaughter
premises. Using the natural movement patterns of the species concerned, sheep, cattle and horses can all be trained to lead. In herds of dairy
cattle, the movement order to milking is quite consistent though the
rear animals have more fixed positions than the 'leaders'. The 'milking
order' is not necessarily the same as the leader–follower order when
moving between grazing areas. Under free-range conditions the older
grazing stock can transfer to their offspring information about seasonal
pathways, areas of good pasture and watering places if this familial
bond is not disrupted before weaning. In this way, home range areas
can be established efficiently. Sheep in extensive pastures may establish
separate home range areas. Subgroups of the whole flock can often
graze with minimal overlap in these regions. In sheep there is a
tendency for the ewe to withdraw from the main flock for up to three
days after giving birth before leading her lamb back into the group.
Imprinting and the neonatal bond are then well established. Where
possibilities do not exist for this withdrawal, other ewes a few hours off
lambing may steal new lambs. Dominant cows can take calves from
subordinate cows, but generally lose interest in them after producing
their own young. The uninterrupted formation of the maternal–
newborn bond, within the first few hours after birth, maximizes survival
potential under free-range conditions. Bucket-rearing of calves in
competitive association can influence later adult behaviour. Rapid
drinking of milk from the bucket may lead to continuation of the
sucking drive, leading to sucking of ears, testicles or other parts of
associating calves, and establish undesirable habits of inter-sucking in
adult stock.

ASSOCIATIVE BEHAVIOUR OF CHICKENS

The chick shows early social responses while still in the shell: it may give low-pitched distress calls if cooled or rapid twitterings of contentment if warmed. Chicks which are hatched at slightly subnormal temperatures give distress calls as their moist down dries and they lose contact with the egg shell. Contact with a broody hen or other warm object prevents these calls. Newly hatched chicks are attracted to the hen by warmth, contact, clucking and body movements. This attraction is greatest on the day of hatching. They develop the behaviours of maintenance, in particular to eat, roost, drink and avoid enemies, in the company of their mother. In chicks the most sensitive period for imprinting is between 9 and 20 hours after hatching. The attachment to the mother is further strengthened as her voice and appearance are recognized. As the down starts to disappear from their heads, the hen rejects the chicks by pecking at them and the clutch becomes dispersed. The clutch is the basis of flock organization and even after it has dispersed chickens need company. A chick reared in isolation tends to stay apart from the flock. Flock birds eat more than single birds.

The peck order is most clearly seen in competition for food or mates and subordinate hens may obtain so little food that they lay fewer eggs. Birds in a flock kept in a state of social disorganization by the removal and replacement of birds eat less and may lose weight or grow poorly. They also tend to lay fewer eggs than birds in stable flocks. Additional feed and water troughs distributed about the pen allow subordinate hens to feed unmolested and an adequate number of nesting boxes ensures that these subordinate birds have continuous opportunity to lay. Flocks of over 80 birds tend to separate into two distinct groups and at least two separate peck orders can be established.

Adult flock formation depends on tolerant association. Strange birds are initially attacked and are only gradually integrated into the flock. Newcomers are relegated to positions near the bottom of the peck order and only active fighting will change it. Hens and cocks have separate peck orders as males in the breeding season do not peck hens.

Debeaking does not eliminate aggressive pecking entirely or prevent the development of the peck order. Pecking by debeaked birds, however, is often ignored by subordinates. Hens kept crowded on wire cannot exercise their normal pecking drive. They often attack other birds and feather-picking may develop.

16   Body Care

Care of the body, through skin hygiene, evacuation and the practice of thermoregulatory tactics, is an on-going component system of the self-maintaining behavioural complex in farm animals. Acts of body care, such as scratching, shaking and licking, are usually brief and varied in form; as a result they are not very conspicuous as a system. But these acts are important and numerous and their total occurrences, per day, constitute a significant proportion of the work of maintenance. Another feature of this behaviour is the flexibility of its acts, allowing it to intrude into other major behaviours such as feeding and resting. Body care evidently has a high priority ranking.

The main divisions of this behavioural class are manifestly related to hygiene and thermoregulation, but other needs also are met by this system. In general comfort is sought and discomfort is avoided. Open sunlit places are used freely and preferentially by horses and cattle in mild temperatures (e.g. about 23°C); shade from direct sunlight, however, is sought in hot temperatures (e.g. over 28°C) by most European breeds of cattle. Shorn and unshorn sheep in early summer will variously seek shade from direct sunshine and shelter from wind according to fleece cover and ambient temperatures. Sheltering and orientation from rain and snow, in cattle and horses, are particularly noticeable when the precipitation is wind-driven (Fig. 55). Shelter from flies is often difficult for grazing animals to achieve unless they can find high ground receiving air currents where they can take a stance. Given a fly problem, cattle and horses may gather closely together and with tail switching they can set up a fairly proficient fly screen for themselves. Agitation of head and ears also contributes to fly defence. Cattle may shake ventral folds of neck skin and horses usually shake manes and forelocks as additional means of creating air and fly disturbance. In the presence of very dense fly populations, cattle will be more likely to stand

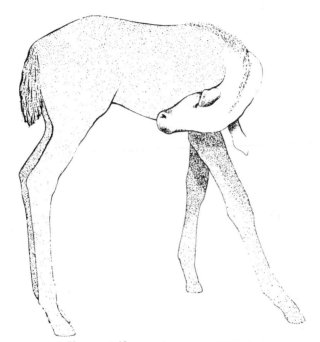

Fig. 54. Self-grooming in a young foal.

Fig. 55. Cow and calf in a snow storm showing heat-conserving postures and association.

in close groups with their heads together. Occasionally they lie with their underparts—belly, brisket, neck and throat—on the ground. This 'grounding' action seals off many sensitive skin areas from exposure to fly irritation.

The storage organs of rectum and bladder are normally emptied when they become distended and postures of defaecation and micturition are adopted in ways which limit soiling of the hind limbs and tail.

The simple assumption can be reasonably made that scratching in animals mediates between a condition of itch and relief of itch and that comfort is obtained by the activity. This is probably one of the most obvious of common acts of body care in animals and serves to indicate the likely nature of the drive, basic to body care behaviour in general. The presence of such a drive is likely to be the result of the needs of body surface protection and thermal homeostasis, to which reference has been made in some examples above. It can be seen, once again, that multiple needs can have a common denominator in the way that they may be served by one convenient, general drive. In this case the unitary motive force, which could be termed a 'physical comfort drive', apparently functions as a caretaker organization. As with other drives, the physical comfort or body care drive is subject to variation with the health and welfare status of the animal. In ill-health generally, the body care drive diminishes and grooming activities become reduced or arrested. In many illnesses the coats of affected animals lose their normal clean and orderly appearance. Such coats, lacking the effects of friction from rubbing, moisturizing from licking, removal of debris from scratching and brisk shaking, become 'staring' or 'harsh' in appearance. Sick animals, with reduced body care motivation, may not discriminate in choice of resting place or time and place of evacuation. Since they are probably also lying for longer periods than usual, their coats may become heavily soiled, particularly about the hindquarters. Similar deterioration in coat condition is also the result of chronic installation or enclosure with inadequate bedding, circumstances which defy operation of natural skin hygiene methods.

Apart from frustration of the body care drive, it may also become altered through other circumstances. Excessive self-grooming occurs in young calves which are subject to acute restraint. If the restraint prevents the operation of other major drives, hypertrophy of a permissible drive, such as body care, may result in the generation of abnormally excessive grooming. Anomalous self-licking has been reported to be comparatively common in some systems of intensive calf rearing which involve isolation and acute restraint of the animal. This behaviour can result in the formation of hair-balls in the alimentary canal. Critical clinical consequences can result from this.

The behaviour of grooming has certain characteristics common to most species. Scratching about the head parts with a hind foot is one. Licking certain accessible parts is another. There are few body areas which are not scratched or groomed in this way by horses, cattle and pigs. Vigorous body scraping in sheep, however, is sometimes a sign of the neurological disease of scrapie.

Other grooming activities are peculiar to the species. Horned cattle frequently rub their horns and horn bases against suitable solid structures. The horn base is an area which can accumulate cornified and epithelial debris. This can gather in the pit of the poll region unless removed. Rolling in horses is one form of skin attention not seen in other farm animals. When the horse is about to roll it sets itself down, with some care, on a selected spot of ground. The animal rolls onto its side and rubs its body onto the ground surface. This rubbing lasts a few moments and the horse rotates towards sternal recumbency, from which position it rolls once more onto its side to rub again. This process is usually repeated and in some of these rolls the horse usually rotates onto its back, holding this position long enough to twist its back once or twice, working the skin of the whole of its back onto the ground. From this position the horse usually rolls back to the starting position but occasionally the entire rolling episode may terminate with the horse rolling from the supine position onto its other side, thereby going through 180° of rotation. At the conclusion of rolling the horse stands and carries out very vigorous shaking of the whole body. Each shake begins at the anterior and passes down the body to the hind limbs. During this shaking the animal's entire hide ripples and dislodges debris, including dust picked up in rolling. This skin-rippling effect of shaking is a phenomenal feature of grooming. It cannot be appreciated by naked-eye observation. The process is revealed by fast cinemato-graphy and slow-motion projection. The rippling creates a surprising looseness of skin. The shedding of debris is also quite remarkable as a result of such a simple, natural and brief session of body shaking. The explanation of the efficiency of this grooming lies, to some extent, in the vigour with which each of the steps is executed. The actual work involved in body care is well illustrated by this behaviour.

Another form of equine grooming is nipping of accessible body areas with sharp, repetitive biting actions. Mutual grooming between horses in pairs is common. The horses face towards each other and nip areas of the other's backs not accessible to themselves, usually behind the withers. This behaviour can be sustained for many minutes during which each animal is continuously nipping actively at the same rate, though not necessarily in rhythm. Horses groom their crests by rubbing them to and fro beneath a manger, tree branch, etc. Manes can become damaged in this way under some circumstances. Yet another form of grooming, peculiarly equine, is scrubbing of the buttocks. This is another region which cannot be attended to by nipping or rolling. A swaying action of the rump is used against a convenient structure such as a post, tree, building, gate or fence. Special scrubbing places become adopted and fences can be broken down by such continual use. While this behaviour can be a sign of parasitism in the horse, it is also normal grooming and has no clinical relevance unless the incidence of the action becomes significant. The more normal effect of this grooming is two-fold. It scrubs the skin of the buttocks and the outer face of the tail head, removing scurf. It also causes the tail to be pressed into and across the skin of the dock, wiping this hairless region which can accumulate skin scales, salt from sweat, sebaceous grease and small faecal accumulations.

The eye, face, nose and nostrils of the horse receive hygienic attention by the animal rubbing its face up and down the side of the appropriate foreleg, which may be held out in front of the other to be more accessible. Horses do not use their tongues to clean out their nostrils as do cattle, but they snort to do so. Nasal secretion in the horse can be considerable. In severely cold weather these secretions can freeze and become very obvious on the horse's muzzle. Abnormal nasal discharge may also accumulate in some illnesses. This is partly due to their excessive production and partly due to the fact that the body care drive is suppressed in most illnesses.

Much body care relates to thermoregulation, as has been stated previously. Some notable examples of this are included in the following illustrations. Swine running outdoors in hot climates have a predilection to wallow in mud. Sheep shelter below ledges in the terrain of highlands and moorlands in high wind, driving snow and hot summer days. Cattle shelter in close groups beneath shade trees during hot hours of the day in the tropics. Piglets huddle in tightly-knit ranks in cool environments. Cattle have a predilection to stand with their feet and lower limbs in water, for long spells in warm sunny days (Fig. 56). Cattle

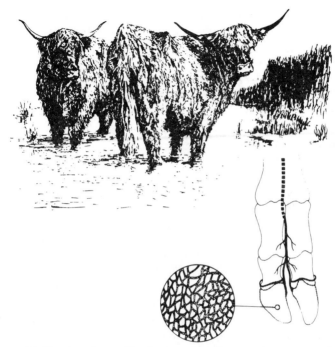

Fig. 56. Highland cattle standing in a pond cooling on a hot summer day. The complex vascularity in the foot region, which makes this behaviour functional, is shown in the inset.

turn their backs into driving rain or snow, closing their inguinal region by adducting their hind limbs closely below them. Horses turn their hindquarters into high winds. Poultry seek shade from hot direct solar radiation and may crowd dangerously in doing so, if the shade area is restricted. Poultry stand with their wings held out from the body when heated. Out-wintering livestock make strategic use of tree clumps, woods, dykes, buildings, etc. as leeward areas, affording protection from chilling winds.

In circumstances of communal sheltering the personal space normally maintained by the group members, as a fairly constant feature of their social organization, usually becomes surrendered. This facilitates huddling, a behavioural arrangement which, in itself, is a proficient thermoregulatory tactic. It serves as a reminder of the overruling priority of the body care drive in critical times of metabolic conservation.

17 Territorial Behaviour

Spatial factors influence many activities of animals, particularly the pastoral ones. In the maintenance, use and negotiation of space the animal meets basic physical and ecological needs. Spacing for animals falls into two general types. Personal space is portable and is therefore carried about with the individual. This type is important for spacing among conspecifics. Actual territory, or the enclosed environment for the animal which may be fixed by a physical boundary, is the other general type. Personal space meets the need for individual freedom and territorial space meets ecological needs. Both involve social negotiations. In territorial behaviour fairly precise rules of conduct determine the tenure of space and dominance privileges within it. Social behaviour is a determinant factor in territorial possession.

TYPES OF SPACE

Territory

Various forms of territory, used by the animal for ecological support, can be recognized. The *home range* is the area which the animal learns thoroughly and which it habitually occupies and patrols. The animal acquires a close territorial bond with the home range. In some cases the home range may be the animal's total range. Within a home range, such as an extensive area of pasture, there may be a *core area*. This core is the area of heaviest regular use within the home range. The demarcation of a core area may not be too precise but generally includes resting areas. *Basal territory* is area in which an animal, as part of a group, will use comprehensively. It may be a fixed geo-

graphical region though it may lack permanence. In any relocation of
an animal geographically, the new territory, which the animal would
require and would seek, would be the one necessary for the requirements of nutrition, shelter, resting, watering, exercise, evacuation,
periodic movement and defensive shifting.

Personal Space

Personal space, which animals seek and preserve through their
behaviour, is intangible in nature and has several forms. One is the
physical space which the animal requires to occupy for its basic

Fig. 57. A cow resting on a strategic crown of land on free range.

movements of lying, rising, standing, stretching and scratching. This
space is somewhat expanded in the head region to accommodate the
greater amount of movement of the head in the course of ingestion,
grooming and gesturing. *Social space* is additional to this; it is the
minimal distance that an animal routinely keeps between itself and
other members of the same species. Each species has a characteristic
minimum distance for tolerating the presence of the nearest neighbour.
This spatial arrangement within social systems is a form of portable
space. This individual or social distance is also personal space. Beyond
the area of social space is the flight distance perimeter. *Flight distance* is
the third form of personal space. It is the minimal distance which an
animal will allow an alien, and a potential predator, to approach before
taking flight. The form of the flight response may be dramatic escape
action or an inconspicuous short retreat. Flight distances become

modified in farm animals when they have become accustomed to managemental manipulations.

The acquisition and preservation of territorial space, as well as personal space, is important in the establishment of dominance. Hierarchial systems seem to be largely concerned with the assertion of one member of a group over another in acquiring prior claim to space which contains preferred characteristics. It is also involved in acquiring preferential access to food, shelter and sleeping areas.

When spatial behaviour in animals is considered it becomes evident that the many forms of space, to which reference has been made above, become integrated so as to provide a mixture of circumstances involving fairly continuous mediation of spatial needs, defence of personal space and spatial surrender. Aggressive activities are, of necessity, common in territorialism. Territoriality, with its aggressive components, can be seen to be the use of aggression for defensive purposes to ensure access to physical provisions and to preserve bodily defence. Maintenance of territories by aggressive behaviour has been widely studied in various species of free-living animals and the laws of territoriality in these studies have become evident. Vocalization, for example, is frequently used in supporting or announcing territorial claims. Various forms of 'sign-posting' activities at strategic points within the home range illustrate yet another technique. Advertisement of 'self' through threat displays is another behavioural feature in territorialism.

SPATIAL NEEDS

Spatial needs in farm animals are both quantitative and qualitative. Quantitative needs relate to space occupation, social distance, flight distance and actual territory. Qualitative needs relate to space-dependent activities such as eating, body care, exploration, kinetics and social behaviour. Both types of needs coexist.

The minimal spatial need is for that amount of room which physical size and basic kinesis require. Even the installed animal, for example, needs distances of length, breadth and height in which to stand, lie and articulate its major parts, including head, neck and limbs. In the acts of rising and lowering the body, the animal is required to make forward and backward movements. During these, the weight, or the centre of gravity, is shifted strategically to or from the fore quarters or hind quarters. The animal uses the weight either as a counterbalance in rising or as a direct pull in lying. Another need for length is in stretching the head and neck forward and one or other hind limb backward. Lateral articulation of the head and neck is still another frequent and necessary form of movement. In lying the animal's need for lateral space is increased through partial rolling of the body and the extension of hind limbs.

Further to physical space, there is an extension of personal space into that invisible but recognizable room which the animal maintains in many social situations. This space is used to keep some separation between itself and others of its kind, its conspecifics. In ranging animals this space is continuously adjusted by changing physical relationships. It

Fig. 58. Aggressive territorial behaviour in a young bull at pasture with breeding heifers.

is preserved in many instances by gestures of threat or intention. Among the farm animals this space is often subject to compression, through enforced grouping. It appears that the last portion of this personal space to be surrendered under such circumstances is a 'bubble' of space about the head. Destruction, or long-term loss, of this head space leads to futile negotiation of interpersonal spatial needs and increases the incidence of aggressive exchanges within an enclosed group. In cattle this head space appears to be about 1 m in radius at its optimum. As with other spatial needs, personal space, including head space, can be surrendered without manifest aggression in a variety of circumstances such as short-term crowding and resting.

The farm animals are 'contact' species; they allow fairly close physical proximity between themselves, except in special circumstances related to sexual, maternal and aggressive behaviour. The distances they maintain between themselves and other potentially predatory animals is much greater. This flight distance is the radius of space within which the animal will not voluntarily permit the intrusion of man, for example. In domestication, of course, this distance shrinks with habituation and socialization. In some horses or pet livestock it may disappear, but most

Fig. 59. Flight response in a goat penned in isolation following intrusion of its 'personal space'.

farm animals retain some flight distance as a spatial need. Sudden intrusion within this may cause an unexpected defensive reaction from the animal. Reactions to this form of spatial loss include startle, alarm, fight-or-flight display and vocalization.

Territory of tangible or actual form is another spatial need, of a quantitative nature. Prior to domestication, the ancestors of the farm animals required a given amount of land for their support. Their habitat had to provide adequate food, safety and shelter for their maintenance and reproduction. When the individual could claim sufficient territory then the size of its group, or the population density, could be set at a

Fig. 60. A Pictish stone carving (8th century A.D.) discovered in north-east Scotland, indicating clearly the typical threat or 'fight-or-flight' display of the bull.

level appropriate to the current carrying capacity of the land. Territory probably then became basically a substitute goal, representing the food it could provide. In evolution, property qualifications have tended to become themselves replaced by the abstract qualification of social status, around which social systems revolve. The farm animals function typically in herds or flocks whose group members possess varying values of social status, permitting their integration into the group using common territory. The struggle for status in the social dominance order is really the struggle for territorial property. Territoriality is therefore partially a social phenomenon. The numerous spatial needs provide motivation for these several behaviours and to this extent, therefore, a territorial drive can be recognized.

COMPARATIVE SPACING

A notable form of the territorial drive in spatial behaviour is home range defence. In farming, modified territory may be that area allowed an animal for its daily activities. This equates with its home range. A home range bond becomes established, leading to the animal's preference for this space and its willingness to defend it. Pasture breeding bulls provide a good example. Farm enclosures are restrictive of range and territory and often enforce changes in behaviour, particularly as population density increases. With confinement, there is increasing impoverishment of behaviour with a reduction in available space. Even in large enclosures, clumping of social subordinates occurs leading to even greater spatial need. Available space and density are separate factors, as can be seen in sheep herds, for example. The social pressures which result from violated personal space appear to be the most important stress factors in crowding. Territorial behaviour continues, in well adjusted form, in high population densities, given adequate space. When space is deficient, however, social behaviour becomes disrupted, even if the group size is proportionally reduced. Although levels of aggression tend to rise in dense populations, the effect of density on aggression is complex. At some levels of space and of population, aggression may decrease with density. It appears that mobility is important since it provides an alternative to competitive encounters in spatial behaviour. Crowding can then be seen to be a stressful condition when there is both a severe spatial restriction and large numbers per enclosed group.

Corners are important features of enclosures and apparently increase the conceptual space of an enclosure. The square has more preferred space than the triangle and the triangle more than the circle. Group size, available area, stocking density and shape of the available area are all variables that influence spatial need.

The qualitative spatial needs of animals essentially relate to ecological circumstances. When these needs are well met in farming, through the adquate provision of feed, water and shelter, various space-modifying factors come into effect. The stability of social dominance order allows maximum group size to function in minimal space. Group composition is therefore important. When this is subject to change, spatial needs for the group will tend to increase. The physical features of territory also modify space needs. High levels of variable features in territory supply more potential for population absorption. Variability will probably provide variety in food, watering places, shelter and travel routes for the innate needs for ingestion, body care, rest and kinesis.

The different species have their own needs and forms of territorial behaviour. In territorial aggression, horses fight with their own typical offensive and defensive weapons. They bite, kick out with their hind feet and strike with their fore feet. Typically a horse will defend by kicking with its hind feet when its flight distance is violated. Both hind feet may be kicked out, for example after backing up to the intruder. This double

Fig. 61. A nineteenth century engraving showing a donkey in the act of delivering a 'mule kick'.

kick is delivered directly to the rear without aim. Kicking out defensively, with one hind limb, is sometimes called the mule kick (Fig. 61), although it is used by horse-kind generally. The mule kick is a directed kick with one hind foot. The farm species often share a common territory such as a grazing area, without interspecific aggression. Horses, however, may show aggression to cattle and sheep grazing with them by biting, kicking or chasing them. Horses are not so prone to take flight as is commonly thought, but they do use territory for running, as a group activity. Equine behaviour contains territorial rituals. Stallions pass faeces in specific sites where their dung may become heaped. All horses use certain territorial spots for defaecation. These areas are not grazed. A restricted grazing territory for horses soon

Fig. 62. A nineteenth century engraving showing spatial behaviours of territorial assertion (extension of the head by the dominant animal) and submission (lowering and turning the head in the subordinate animal).

becomes divided up into 'lawns' and 'roughs'. The lawns are the areas closely cropped and the roughs are the dung areas which remain ungrazed, except in starvation conditions.

Sheep and pigs threaten, to obtain or defend personal space, by extending their heads at one another. Failing a submissive response, such as moving away the head, an aggressive action may follow. Pigs will bark and root at each other. Sheep may push one another, butt or tug strands of wool. All species are capable of showing determined defensive actions to hold territory which has become home range. When removed from home range, most livestock show keen efforts to return. In horses this homing is quite a phenomenal behaviour. Cattle also show a desire to return to known places from which they have been moved. Sometimes the home bond is evidently great and is a behavioural feature which has not always received adequate consideration in animal management.

18 Behaviour of Rest and Sleep

All clinically healthy domestic ungulates lie down on the ground at least once a day either to rest or to sleep. They spend a significant proportion of their lives recumbent. Resting and sleeping are governed by timing controls, more obviously than other cyclic activities. The purpose of rest and sleep is restorative, allowing metabolic recoveries and conservation of energies. This system of passive behaviour is of critical importance in the maintenance of the animal and receives high priority in environments to which the animal is adjusted. Sleep occupies much of the lifetime of animals.

SLEEP CONTROL

A biological clock evidently exists that measures time and issues temporal signals to the rest of the body. At the cellular level there appear to be cyclic variations in macromolecules within the cell. Rhythms of intercellular activity also occur. Noticeable rhythmic activity is also a feature of the endocrine system. Of fundamental significance is the fact that levels of deoxyribonucleic acid (DNA) and ribonucleic acid (RNA) show rhythmical diurnal changes. A master clocking system in the body seems to exist which adapts the rhythms of organs and cells to the environmental temporal clues such as photoperiodic stimuli. Cells in the lateral hypothalamus show a diurnal rhythm of responsiveness to stimulation.

Two forms exist: brain sleep (slow wave sleep or quiet sleep), and the sleep of the body (paradoxical or rapid eye movement sleep, REM). Slow wave sleep (SWS) is quiet sleep in which EEGs show a low level of electrical activity in the brain. Paradoxical sleep is called this because of the paradox of manifest sleep in association with electrical activity in the brain typical of wakefulness. Paradoxical sleep is also characterized by rapid eye movement (REM). The two types can best be differentiated from wakefulness and from each other by electroencephalograms. In slow wave sleep, the EEG is characterized by synchronous electrical waves of high voltage and slow activity. In paradoxical sleep the EEG shows low voltage and fast activity similar to that seen in the wakeful state, but there is very little muscular activity and the animal is more difficult to arouse than when it is in SWS. During body sleep, or paradoxical sleep, the muscles of the eyes frequently contract hence the term rapid eye movement (REM) sleep. Humans awakened from REM sleep report that they were dreaming. In animals, vocalizations, leg movements and eye movement, of piglet sleep for example, indicate that animals may also experience dreaming. REM sleep appears to be the critical or necessary component of sleep since REM sleep deprivation results in behavioural abnormalities in all species tested. The nucleus of the hypothalamus is currently thought to be the likely locus of the biological clock. Major circadian rhythms are under the control of this region. Light stimuli are transmitted via the optic nerve ultimately to the pineal gland which is evidently an intermediary unit in the synchronization of diurnal rhythms of behaviour. Signals from the pineal body may travel to the medial forebrain and the reticular formation of the spinal cord, thus influencing the behavioural predisposition of the organism. The hormone melatonin, produced by the pineal, is present in higher quantities in plasma at night. Melatonin levels also appear to be the link between the photoperiod and annual cycles of sexual activity. Day length influences the relative amounts of both serotonin, a precursor to melatonin, and melatonin present in the pineal. These influence the organism's resting behaviour.

The build-up of serotonin may induce drowsiness and sleep. This humoral basis of sleep is further supported by studies in which cerebrospinal fluid of sleep-deprived goats induced sleep in rats. Clinically, tryptophan, a precursor of serotonin, may be used to induce sleep. Acetylcholine has been implicated as the neurotransmitter responsible for wakefulness and norepinephrine that for REM sleep.

REST AND SLEEP IN FARM ANIMALS

Horses are polyphasic animals as regards sleep or rest periods; 95% of horses have two or more such periods per day. The total length of time

Fig. 63. Normal sleeping posture of the adult horse. The upper limbs are placed away from the trunk, in front of the fore and behind the hind lower limbs.

Fig. 64. The resting postures of the horse as seen in one group of three: fixed stance (rest), sternal recombency (drowse), lateral recumbency (sleep).

spent recumbent per day is approximately two and a half hours, with slight variations associated with age and management. Twice as much time is spent in sternal recumbency as is spent in lateral recumbency and normal adult horses rarely spend more than 30 minutes continuously in lateral recumbency; the mean time spent continuously in this position is 23 minutes.

Horses lie down and rise in a specific manner. In descent the forelegs are flexed first, then all four limbs. The head and neck are used for balance as the animal lowers itself. In sternal recumbency, horses do not lie symmetrically. Their hindquarters are rotated with the lateral surface of the bottom limb on the ground. In lateral recumbency the upper limb is invariably anterior to the lower forelimb which is usually flexed. The hindlimbs are usually extended with the upper limb slightly posterior to the lower limb. To rise the horse extends the upper foreleg. The forequarters are raised so that the lower foreleg can be extended. At the same time both hind limbs begin to extend, but the main thrust of raising comes from the hind limbs.

A sound horse seldom remains lying when it is approached, probably because a standing horse is better able to flee or to defend itself. It is interesting that a dominant stallion in a herd lies down first, before subordinate horses. Horses may be able to drowse and even engage in slow wave sleep while standing by means of the unique stay apparatus of the equine hind limbs. In REM sleep, however, they almost always are lying down.

During the day the horse is awake and alert over 80% of the time. At night, the horse is awake 60% of the time, but drowses for 20% of the night in several separate periods. Stabled horses are recumbent for two hours per day in four or five periods. Ponies are recumbent for five hours per day and donkeys rest even more. Slow wave sleep occupies two hours per day and REM sleep occurs in nine periods of five minutes duration.

Management practices can affect equine sleep patterns. When moved from stable to pasture, they do not usually lie down during the first night and total sleep time remains low for a month. If horses are tied too short in a stall, so that they cannot lie, they do not have REM sleep. Horses stabled only at night may limit sleep until they are free during the day. Care must be taken not to deprive horses of sleep as may occur when they are transported long distances and must be tied short in stalls for support. Diet also affects sleep time in horses, as it does in ruminants. If oats are substituted for hay, total recumbency time increases; fasting has the same effect.

During the day cattle usually rest in sternal recumbency while ruminating (Fig. 65). About five hours per day may be occupied in this fashion. They may also lie at rest without ruminating. While not actively grazing, cattle may also spend time standing at rest without ruminating. This 'loafing' time is variable but is short in healthy cattle. In daylight, a total time of less than one hour is likely to be spent in lateral recumbency and episodes of rest in this position are normally brief. These may be associated with periods of sleep.

The presence or absence of true sleep in ruminants has been controversial till recently. The extensive studies of sleep carried out in recent years indicate that cattle show both REM and slow wave sleep. REM sleep occurs in about eleven periods so that the total of about 45

Fig. 65. A herd of Ayrshire heifers resting together in sternal recumbency.

minutes of REM sleep and three to four hours of SWS are divided into many short naps. When cattle are REM sleeping, they usually are lying down with their heads turned back into their flank. Cows that have not yet adjusted to stanchioning are sleep-deprived. Most characteristic of ruminant rest are the extensive periods of drowsiness usually associated

Table 5. *Hours per day spent awake, drowsing and sleeping in horses, cattle, sheep and pigs*

Animal	Diurnal period	Awake	Drowsing	Sleeping
Horses	Day-time	12·9	0·9	0·6
	Night-time	5·3	1·5	2·8
	Total	18·2	2·4	3·4
Cattle	Day-time	10·6	1·2	0·2
	Night-time	1·9	6·3	3·8
	Total	12·5	7·5	4·0
Sheep	Day-time	10·0	1·6	0·6
	Night-time	5·9	2·7	3·2
	Total	15·9	4·3	3·8
Pigs	Day-time	7·4	2·5	2·0
	Night-time	4·4	2·5	5·2
	Total	11·8	5·0	7·2

After Ruckebusch (1971).

with rumination. At this time they usually lie on their sternums. Cattle are in a drowsy state for about seven to eight hours per day divided into 20 periods or more that precede and follow sleep. Rumination and sleep are inversely related so that sleep time decreases with alimentary development. Sternal resting increases but sleep decreases as the percentage of roughage in the diet increases.

Normal behaviour patterns of sleep, or hypnograms, have been used as an indication of stress-free husbandry. Disinclination to lie at rest is seen in horses with orthopaedic conditions and in cattle with hardware disease.

Farm personnel should recognize the normal sleep and resting patterns of animals so that abnormalities which may have symptomatic significance can be detected. A horse lying down at night in a normal posture is probably asleep. An adult horse that lies down during the day, unless in the company of its foal, or rests in sunshine may be abnormal and should be carefully observed for other evidence of illness. Although cattle do not sleep much, they may be difficult to arouse when they are asleep at night. Pigs lie and sleep a lot so that protracted resting and sleep are positive indices of normal function. Some clinical conditions cause disturbed patterns of resting and sleep in all animals. Significant clinical signs include sleeping fitfully and rising from resting postures frequently.

Management practices should interfere as little as possible with normal circadian patterns of maintenance behaviour. Interruption of activity cycles and loss of sleep may play an important role in the aetiology of the stress-related diseases associated with newborn management, livestock transport, mixing of strange animals and introducing new animals into established groups.

Part IV

Species Behaviour Patterns

19 Behaviour Patterns in Horses

REACTIVITY

Horses are unpredictable in the way in which they display agonistic behaviour. Under farming conditions, the response to alarm or threat may be flight or attempted flight on the one hand or attack on the other. Horses attack using their teeth and hind feet. It is thought that aggressive acts in horses maintained in isolation may be the result of excitement or habit. Many stockmen and experienced horsemen hold the view that a tendency towards aggressive behaviour in horses can be recognized in association with certain physical features. They also hold that certain colours in horses and certain characteristics of the head and the eyes portray a temperament in which aggressive behaviour is a feature. While it is difficult to refute opinions based on long experience by association, it must also be recognized that such opinions may be based on prejudice. It is certainly difficult to justify, on scientific grounds, belief that a horse's temperament is predetermined by certain morphological characteristics.

Vocalization

It has been suggested that the horse makes three basic sounds: a neigh, a grunt and a high-pitched crying noise. These sounds vary in their degree of intensity and duration and also show variations according to sex and age and the particular stimulus which elicits them. There are also specific sounds which are variations of the three basic ones.

The neigh is the loudest noise emitted by the horse. It is often heard when a mare is separated from her foal or when a horse is curious about events outside its range of vision, or when it is seeking to communicate with other horses. Three types of grunt can be distinguished: the most

Fig. 66. Full volume vocalization in a stallion, directed at distant horses.

frequent just prior to feeding, the second emitted by stallions at the beginning of a sexual encounter and the third when the mare has cause to worry about her foal. The crying or squealing sound can vary in volume and is usually heard during aggressive encounters, i.e. as a part of threat behaviour, in fighting or in instances of aggressive sexual rejection.

Additional forms of vocalization include trumpeting in the male and the throaty gurgling of general satisfaction.

INGESTION

Horses graze by cropping the pasture close to the roots with their incisors. Whilst grazing they cover large areas and seldom take more than two mouthfuls before moving at least one step further, avoiding grass patches covered in excreta. They maintain some distance between each other when grazing in groups.

The young foal does not graze very efficiently until it is several weeks old. By about the end of the first week of life, however, the foal has begun to nibble the herbage in association with its dam. Horses have been seen to spend some time vigorously de-barking trees, but the reason for this apparently aberrant form of eating behaviour is as yet unknown and it is possible that some wood eating is normal.

Horses do not drink very frequently in a 24-hour period and many may only drink once a day. When they do drink they typically consume very large quantities of water, totalling up to 15 to 20 swallows.

EXPLORATION

A great deal of exploratory behaviour is shown by the newborn foal. This behaviour is directed towards the pasture, the ground, the premises and their boundaries and other objects in the environment within its touch. In the course of this exploratory activity the foal may nibble and mouth unfamiliar objects. Such keen exploratory behaviour is not shown to the same degree in adult horses, except between themselves. They nevertheless acquire familiarity with allocated territory, and familiar territory is evidently quickly adopted as the home range. Horses use eliminative behaviour to a large extent in defining their territory and the home range becomes marked and mapped out by deposits of excreta. Preference for the home range is very strong among horses and they typically show more willingness to be moved towards home than away from home.

KINESIS

Locomotion in the horse may take one of four different forms. These gaits can be distinguished by the sequences in which the hooves are lifted from the ground and alter with the acceleration of speed from the walk, through the trot, to the canter and thence the gallop, which have been given previously in detail; a brief review, however, is given below.

The canter and the gallop are mechanically identical in their hoof-lifting sequences, but differ in rate of performance or speed, the gallop being an accelerated. canter. At slowest speed, i.e. the walk, the sequence of hoof-lifting proceeds as follows: near hind, near fore, off hind, off fore. This is repeated with a return to near hind. In the trot the animal is balanced alternately on diagonally opposite feet and the sequence of hoof lifting is as follows: near fore with off hind, off fore with near hind. This is repeated with a return to near fore with off hind. In the gallop the horse uses the same action at maximum speed, but at one point there is a floating phase during which all four feet are collected under the body and are off the ground. When the off forelimb throws the horse into this floating phase it is termed an off-leg lead; when the near forelimb throws the horse into the floating phase it is, conversely, a near-leg lead. A typical sequence of feet movements in the gallop is as follows: off fore, floating phase, near hind, off hind, near fore. This sequence is followed by a return to off fore. With a near-leg lead, the sequence is as follows: near fore, floating phase, off hind, near hind, off fore. In the gallop there are never more than two feet on the ground together. In the canter, however, both hind feet are still on the ground when the first forefoot touches the ground.

During locomotion, horses are capable of effecting quite spectacular jumps, both in distance and height covered. Few horses, however, jump often until they are taught to jump and, indeed, untaught horses may avoid obstacles only 60 cm high rather than clear them by jumping. Normally horses avoid jumping over ditches and show a reluctance to jump over horizontal obstacles in general.

ASSOCIATION

Whilst being a typical herd species, horses also show a marked preference for certain individuals of their own species. Two horses encountering each other for the first time show much more mutual exploratory behaviour than is seen in the other farm animal species. Exploratory behaviour at introduction involves an investigation of the other's head, body and hindquarters using the olfactory sense (Fig. 67).

As with the other farm animals, horses show a form of social order when they live in groups and a social hierarchy becomes established

Fig. 67. Exploratory social interaction between strange horses on meeting.

within these groups. The older and larger animals are usually found to be high in the dominance order. Stallions do not necessarily dominate geldings or mares. A dominant individual often dictates the movement of the herd throughout the grazing area and will sometimes break up exchanges between other horses. Socially dominant horses are sometimes found to have more aggressive temperaments than the others. Horses running at pasture show special features of behaviour if the group contains a stallion and breeding mares. Stallions usually drive younger male animals to the perimeter of their groups, but will not show any aggressive attitudes towards them if they remain there. The stallion attempts to herd a group of brood mares together. The normal size of 'harem' amongst horses is about seven to eight mares. The colts

Fig. 68. Mutual grooming in a pair of horses bonded in a social arrangement.

tend to form a bachelor group after splitting off from the herd at the age of about one to two years. Fillies may or may not join this group.

BODY CARE

At pasture, pairs of horses may spend quite lengthy periods in mutual grooming (Fig. 68). This form of grooming is shown in all age groups. Though the pairs which are formed by individuals are usually matched for age and size, mares and their foals often groom each other. In the normal grooming position, two horses face each other; one extends its head past the side of the other's neck and nibbles vigorously over the latter's saddle region. After engaging in this nibbling action, the initiating animal is usually soon being groomed by the second one.

Horses also spend time grooming their own bodies, particularly around the hip and flank. This is effected by turning round, extending the head and nibbling repeatedly at the skin in these regions. They may occasionally also groom their limbs, both fore and hind, in the same manner.

Rolling at pasture is also a form of grooming in the horse. Here a horse lies down in a normal fashion and then proceeds to kick itself over on to its back and rubs its back against the ground while keeping its feet up in the air. After several rubs it rolls on to its side again. It may roll on to its other side, while at other times it rolls back on to the side from which the roll started.

Elimination

While horses are emptying their excreta from the storage organs of bladder and rectum, they usually cease other body activities. Stallions show careful and deliberate selection of the spot where defaecation is to occur. Following defaecation, a stallion usually turns and smells the spot where it has taken place. After defaecation, in the case of both the stallion and the mare, the muscles of the perineum contract and the tail is lashed downwards several times.

While urinating the stallion and the gelding adopt a characteristic stance, the hind-legs being abducted and extended so that the back becomes hollowed. Urination takes place with the penis released from the sheath. Following urination, the stallion again smells around the area before walking away.

The mare, when urinating, does not show the same marked straddling posture as is shown by the stallion; nevertheless, the posture is similar in that the hind-legs are abducted from each other. Following urination by the mare, the vulvar muscles contract. More elaborate patterns of urination are shown by brood mares with young foals and by mares in estrus.

As already mentioned, horses typically show care in selecting areas for defaecation. They return again and again to the same patch. These patches can accumulate large quantities of faeces during a grazing season. Adult animals defaecate six to twelve times per day, depending on the nature of the feedstuffs eaten. Normally urination occurs less often during the day and horses have been noted to urinate as few as three times per day. Most urine is passed during rest periods in the hours of darkness.

TERRITORIALISM

Free-ranging horses spend much of their time grazing, about 12 hours or more being common. In relatively arid areas horses often travel very long distances daily to water. Their grazing range, under these conditions, is probably limited by the availability of water. When snow is on the ground horses can obtain fluid by eating it and are then independent of running water and can utilize different territories. In free-ranging horses, trips to water and to salt each day are a territorial requirement.

REST

When the horse lies down all the legs are gathered together under the body, the knees and hocks are bent and the chest and forequarters make contact with the ground before the hindquarters. The adult horse normally rests slightly on one side of its chest with one foreleg and one hind-leg underneath the body. On rising, the horse stretches out both forelegs, raising first the forequarters and then the remainder of the body on its hind feet. Adult horses do not lie for very long periods. Mares with young foals tend to lie longer than usual when the foal is near by and sleeping in full lateral recumbency. Mature horses are unable to lie in this flat-out posture for long periods of time before their respiratory functions become impaired. The full weight on the thorax of the horse, when laid flat, appears to be such that circulation to the lungs becomes inefficient after about 15 minutes. This is not the case among foals and young horses, however, and these subjects can be seen to spend many hours in the day sleeping on their sides at full stretch.

Many horses accumulate six to seven hours of rest during each 24-hour period. Some of this sleep is accumulated during the hours of daylight and is largely achieved in the standing position. Typically, periods of sleep are short and irregularly spaced with rest. No regular patterns of resting or sleeping have been observed in adult horses.

It is unusual to see all the members of a group of horses lying down simultaneously. Knowledge of normal lying and resting behaviour in horses was inadequate in the past and led to some unsatisfactory methods of securing animals in recumbent positions for surgical intervention. For example, a horse naturally never lies on its back except when grooming itself, but when placed in this position for surgery it soon suffers from the effects of pulmonary stasis. Again, when a horse is cast and subsequently rolled in order to place it for the surgeon, it is not always remembered that when the horse rolls naturally it normally pauses in midroll with all four feet in the air momentarily. Quick or hasty rolling, especially in large animals or those with full intestines, can create a gut twist.

20 Behaviour Patterns in Cattle

REACTIVITY

When one bovine animal makes a passive approach towards another of the same species, a mild threat by the latter may often be enough to discourage the approaching animal from engaging in physical contact. If, however, the attacked animal is slow to react it may be butted, often from the rear. The nature of the up-swinging motion of the butt may cause serious injury to a bull or cow, particularly if the attacking animal has horns.

An active approach on the other hand, by one animal towards another, is seen when the former makes a deliberate threat. If the latter animal resents this approach, its resentment is indicated by the lowering of its head as in aggressive behaviour, but the animal's forehead is positioned nearly parallel to the ground with its neck extended. However, if the animal being threatened in its turn displays threatening behaviour, fighting ensues. In some cases the two opponents stand a few metres apart with their heads lowered, hind legs drawn forward, eyes on each other and with their horns directed in the same manner. The threat posture of females is similar to the fight or flight posture of males. Other forms of threat behaviour are seen when an animal paws at the ground, rubs its head and neck on the ground and also its horns, when these are present.

When fighting ensues, the animals fight with their heads and horns and try to butt each other's flanks. If one animal manoeuvres itself into a position whereby it can butt the flank of the other, the second animal turns round to defend itself and attempts a similar attack. Fighting does not normally last longer than a few minutes, but in cases where the animals are equally matched the 'clinch' move may be employed. This is a move where the animal being attacked from the side turns itself parallel to the other and pushes its head and horns into the region of the other's udder. This often results in a temporary cessation of fighting which may last for several minutes before action is resumed. When one animal submits it turns and runs from the other which may assert its dominance by chasing after it for a few metres. If neither animal submits, fighting may continue until both opponents suffer from physical exhaustion.

Fighting between females, apart from causing physical injury, can result in a reduction of milk-yield since the inhibited subjects may not feed properly in a restricted area. It is best to keep the new cows in a field adjacent to that of the main herd at first, before introducing the animals one by one to the other members, thus allowing the individuals to become used to each other. Even then fighting may occur as the new animals may struggle to attain positions in the social and domination hierarchy of the whole herd.

INGESTION

Cattle have to rely for food intake on the high mobility of the tongue, which is used to encircle a patch of grass and then to draw it into the mouth, where the lower teeth and the tongue are used to sever the bound grass.

The nature of a cow's feeding accessories is such that it is virtually impossible for the animal to graze less than 1 cm from the ground. When grazing, the cow moves about chewing the grass or plant only two or three times whilst, at the same time, moving its head from one side to the other seeking the next patch of herbage to feed on. In this manner cattle graze mostly during the hours of daylight and cover, on average, about 4 km per day. The distance travelled increases if the weather is hot or wet or if there is an abundance of flies around. During the season of hot weather, more grazing may be done at night than during the day. In each 24-hour period, there are four main periods of high ingestive intake: (1) shortly prior to sunrise, (2) mid-morning, (3) early afternoon and (4) near dusk. Of these distinct periods, the hours prior to sunrise and around dusk appear to be the periods of longest and most continuous grazing. During other times of the day, cattle graze intermittently and idle, rest or ruminate. At such times selectivity is low; prior to the periods of high intake, selectivity of pasture and food plants increases, becoming very marked during the high intake phase. After a while grazing becomes intermittent again and the level of selectivity decreases.

Newborn calves do not graze until they are several days old and their first attempts are usually inefficient. As the periods of suckling are reduced, grazing becomes more regular and calves become highly selective in their grazing intake.

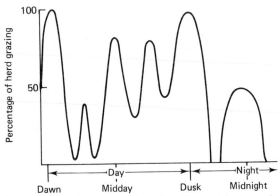

Fig. 69. A typical daily pattern of grazing activity in cattle.

The average time cattle spend grazing during the 24-hour period is four to nine hours (Fig. 69). The periods of rumination may also total four to nine hours and the number of ruminating periods may be 15 to 20. The number of drinks taken per day is one to four and the hours spent lying down are usually in the 9- to 12-hour range. These figures may vary in different respects between beef cattle and dairy cattle and in tropical and in free-ranging herds, but behaviour in domestic cattle is usually fairly stable and the figures given are general norms.

Cattle, like sheep, show evidence of conscious selective intake when grazing, and the same controversial theories of 'nutritional wisdom'—as to whether an individual animal is able to balance its metabolism with the correct amount of the various necessary constituents—have been propounded. Cattle show a particular tendency for selection of certain plants and grasses. Their degree of selectivity can be high and they may prefer plants of a particular species, and even those species only at a certain stage of growth. The effects of this selectivity can be seen in the condition of a pasture area inhabited by cattle. The growth on the grazing area is erratic; while some places may have thick growths of herbage, others are thin and patchy indicating the existence of a favourite food. Cattle usually avoid plants which have been contaminated by excreta or have a hairy or coarse appearance. They may select different types of grass by smell, but the stimulus of taste is thought to be the main factor in deciding selectivity when grazing. The stimuli of sight and smell do both influence the pattern of the grazing behaviour of cattle, but they play only a minor part in food selection. Calves are able to consume a fairly large quantity of feedstuffs which contain an acid ingredient, but cattle prefer glucose to acid foodstuffs and salty to sour substances.

With regard to their ability to maintain correctly a balance of nutrients in their systems, it has certainly been found that some cattle are able to correct a sodium deficiency; but, as with sheep, experiments have not been extensive or conclusive enough to warrant any singular statement on the subject.

Cattle and sheep, being ruminants, can utilize more fibrous foodstuffs and require greater bulk. In addition cattle show more variety in ingestive behaviour when grazing extensively. In general, large animals eat more quickly than small ones and alfalfa requires more chewing before ingestion than ground corn which, in its turn, requires more chewing than shelled corn. The time taken to consume food varies according to its volume, the concentrates which may be in it, whether it is wet or dry and the way it has been processed before being given to the animals. In dairy cattle a lot of eating takes place shortly after milking. Given a choice between silage and hay, milking cows will spend more time at the silage, often two-thirds of the total eating time, while spending the remaining one-third at hay. In a study of food choice, it was found that green fodders and roots were preferred to protein while cereal chaff was preferred to straw.

In the main, the behaviour pattern of cattle in temperate climates is fairly regular and the individuals characteristically behave as a group, idling, resting, ruminating or grazing at the same time. Cattle have, however, been known suddenly to change their preferences, when supplied with food in an enclosed area over a long period of time.

Rumination

Following ingestion comes rumination. Rumination allows cattle to regurgitate, masticate and then swallow food which they have previously ingested into the rumen. Thus animals can continue their digestive activities at leisure, when away from a preferred grazing area or sheltering during bad weather. Usually cattle prefer to lie down during rumination although in bad weather, e.g. heavy rain, they may prefer to stand or walk about slowly. Rumination does occur in young calves, but only takes up a proportion of daily time comparable to that in adults after about six to eight months of age. During the 24-hour cycle, rumination takes place about 15 to 20 times, but the duration of each period may differ vastly; it may last only two minutes or so or it may continue up to one hour or even more at one stretch. The peak period for rumination is shortly after nightfall; thereafter, it declines steadily until shortly before dawn when grazing begins. Times may differ, however, according to diet; cattle are able to regurgitate, remasticate and reswallow long hay more quickly than ground hay or concentrates. The relation between the time spent grazing and the time spent ruminating varies depending on the season and the abundance and quality of the herbage provided, along with the area available to the cows and the size of the herd. The time spent ruminating amounts, on average, to three-quarters of the time spent grazing. Good quality herbage nearly always shortens the time spent ruminating, while herbage which is rough increases the number and length of ruminating periods. In the spring and autumn the time spent grazing almost totally eclipses the periods of rumination, but in the summer they are almost equal.

The factors which may disturb or cause the cessation of rumination are various. During estrus ruminating nearly always falls away, but it does not stop altogether. Any incident which gives rise to pain, hunger, maternal anxiety or illness affects ruminating activities. The periods before and after parturition are not conducive to rumination and it may decline to a low level. Some animal scientists hold the view that the longer rumination is interrupted or delayed, the more difficult it becomes for the animal to resume this activity.

Drinking

This activity refers to the total consumption of water, including that water which is often contained in the animal feed. Cattle usually drink one to four times a day in temperate climates; they do so more often in hot weather and when there is a high proportion of concentrates in their food.

Cattle drink using their muzzles. The tongue, unlike at grazing or feeding, plays little part in the process and the nostrils are kept above the surface of the water. Cattle usually drink in the forenoon, early afternoon and evening but rarely at night or at dawn. More drinking is done on old pastures than in nutritious grazing areas. Cattle given an abundance of feedstuff—a situation which may occur during housing—tend to consume more water than they would normally. In addition to hot weather and an abundance of various types of feedstuffs, milking also induces cows to drink. Thus soon after milking and especially after the evening milking, cattle drink water whenever possible. It may be significant that milk is 88% water.

Several other factors alter or discourage drinking activities. The water intake increases during later pregnancy and lactation, and the intake varies according to the ambient temperature, breed, age, body size, intake of pasture and the level of nutrient and salt in the food provided. European breeds of cattle drink more than tropical breeds whether during temperate or very hot weather and cattle fed on foodstuffs with a high level of protein drink much more than those on a lower protein supplement. The amount of water consumed by pregnant heifers has been calculated to be 28 to 32 kg per day while the average daily intake of water by adults is about 14 kg.

EXPLORATION

One of the main features of exploratory behaviour is that it is engaged in by an animal only as long as the emotions of fear or apprehension are not present. The animal's curiosity is aroused when it sees an unfamiliar object or hears an unknown noise. What may induce exploratory behaviour in one animal may very often be ignored by another. Older animals, being more acquainted with the objects and sounds of the their environment, are less curious and exploratory behaviour is therefore a character of young animals. When curiosity is first aroused the animal assumes a posture similar to that of surrender or submissiveness, but with nostrils quivering and sniffing. The size and nature of the object, in which the animal has become interested, determines the speed of approach. It sniffs the object and may lick or even, if the object is malleable enough, chew and swallow it. This kind of exploratory behaviour is often induced by the sight of familiar objects in unfamiliar surroundings or *vice versa*. It is stressed, however, that the animal's curiosity is rarely followed through if it has any cause for fear or apprehension.

KINESIS

The characteristic aspects of behaviour displayed in play activities by cattle are prancing, kicking, pawing, snorting, vocalizing and head-shaking. These are seen particularly in young calves, although adults do occasionally indulge in playful activities. It has been said that play behaviour provides the participating animal with its own method of release of energy drive and, with regard to the young, has the purpose of helping the animal to acquire the essential motions of behaviour encountered in adults. Conversely, the theory has been put forward that play is simply an activity for its own sake without any underlying purpose or goal and that, in the case of playful fighting, the participants are not concerned with which will win, but only with participation in the activity itself. Playful fighting is distinct from aggressive interactions in that, while playfully engaged, either animal's attention is easily distracted.

ASSOCIATION

One of the most important patterns of behaviour in the range of group interactions is that of the social hierarchy or dominance order in a group of animals which may spend a long time living together. This is specially notable between cattle. When a social hierarchy is set up in a herd of cattle, it will usually last a very long time, although minor adjustments to an individual's social rank are inevitable. The existence of a solid social hierarchy is important for the welfare of a herd.

The way in which this order is evolved often varies from breed to breed. Chest girth may sometimes be enough to assert one animal's dominance over another; on the other hand, height at the withers or total weight may be the deciding factor. It has been found in closed herds that seniority, i.e. age, often influences the social order. It is worth noting that in dairy herds where seniority defines social rank, no newcomers have been introduced into the herd, the only new animals being the ones that are born within the herd. In large open herds, where animals may encounter each other for the first time at any age, the order is usually based on strength and weight. It has been stated after observations that in general Ayrshires dominate Jerseys and that Angus dominate Shorthorns who in their turn dominate Herefords. In these cases, the assertion of one animal's domination over another is often based not on size, strength or seniority but on the hereditary characteristics of the members of the different breeds.

In cattle varying types of social hierarchy can be seen. There is the *linear hierarchy* in which one animal dominates another, which dominates a third, the hierarchy continuing in this manner through the entire herd, with the last animal dominated by all the others. This social order is not seen markedly in large herds and usually only exists within quite small herds in which the animals have been living together over a long period of time. Linear-tending hierarchies are slightly more complex and occur more frequently. These hierarchies have a normal linear dominance order, except in one sense. At the top there is a situation where one animal dominates all the others, except for one, which in turn is dominated by a third animal which is above all the rest except for the first. Thus there is a triangle at the top of the linear-tending hierarchy where each animal dominates all but one. In linear-tending situations, this triangle may occur in the middle of the social order or at its bottom.

Another social order is the *complex hierarchy* where several animals may dominate a certain number in the group, while being themselves dominated by others, which are dominated by several of the initial animals.

Social hierarchies often change as the young males in a mixed group begin to challenge the females in the group. When they are about 18 months old, they come to dominate all the females and join the other adult males which also dominate the females. Fierce fighting is not specifically required to form the hierarchy, even when a new animal has been introduced. Often a threat posture or a movement of the head is enough to determine one animal's dominance over another and there is evidence that the main structure of the order is based on visual factors. Even in breeds with spectacular horn development, such as Highland cattle, these exchanges are purely formal.

It has often been assumed that the members of the herd which are leaders, by the way they influence the movement of the others, are the most dominant animals in the group, but this is not invariably the case. In a free-moving herd, the animals which occupy the middle region of the dominance order are usually the ones which lead. They are often followed by animals which, although they are at the top of the dominance order, are content to occupy the middle of the moving group while the animals most dominated remain at the rear. When cattle are put in a position where they are required to move, for example, into a milking parlour, the order in which they do so is fairly constant, but seems to bear no relation to the weight, size or age of the animals (except for the fact that pregnant cows are usually found at the rear of any group being moved forcibly). While grazing, one or two animals may lead the rest moving in a particular direction but if these two leaders separate and the herd follows the movement of one leader then the other is forced to concede and to return, taking up the general direction of the herd. Usually the two leaders are able to adjust to each other and to avoid splitting up the herd.

It thus appears that there are three social orders in a group of cattle: the social or dominance hierarchy, the order in forced movement and the leader–follower structure.

BODY CARE

Cattle lick and thereby clean every part of their bodies that they can reach. To groom inaccessible parts they often make use of trees and fences and by using their tails they keep off flies and brush their skins. The value of grooming is seen in that it helps to remove mud, faeces, urine and parasites and thus greatly reduces the risk of disease. It has been estimated that calves spend up to 52 minutes a day grooming themselves by scratching and licking. Adult cattle may lick themselves on 152 occasions during a day and scratch 28 times a day.

When one animal grooms another, it is commonly found that the one engaged in cleaning is slightly below the other in the social order (though normally within three positions). In large mixed herds, adult males will groom each other more often than younger animals or females. Their grooming is applied mostly around the area of the head and neck. It has been suggested that one animal grooms another to enjoy the salt properties of the outer skin layers, but it could also be the behaviour of a subordinate creature appeasing its superior or even indicate motives not yet appreciated.

Elimination

During the 24-hour daily cycle cattle normally urinate about nine times and defaecate twelve to eighteen times. The number of times cattle engage in eliminative behaviour and the volume that is expelled, however, varies with the nature and quantity of food ingested, the ambient temperature and the individual animal itself. Holstein cattle may expel 40 kg of faeces in the 24-hour cycle while Jerseys are found to defaecate some 28 kg under the same conditions.

Although the eliminative behaviour of cattle is neither specifically regulated in the frequency of its occurrence nor consciously directed at a certain area, large amounts of faeces are often placed closely together. At night and during bad weather cattle tend to bunch and this appears to be the only reason for the close deposition of faeces. The animals pay little attention to the faeces, often walking and lying amongst the excreta. There is evidence that in some dairy cows allelomimetic behaviour is engendered and when one animal defaecates or urinates others may commence to do likewise.

The normal defaecation stance for both male and female animals is one in which the tail is flexed away from the posterior region, the back arched and the hind legs placed forward and apart. The posture assumed is such that there is the least possible risk of incurring contamination. This attitude towards hygiene is also seen in calves which in fact take more care than adults to expel the faeces well away from the body. Unlike the female, the male bovine animal is able to walk while urinating and only displays a slight parting of the legs while doing so. The posture assumed by the female while urinating is very much the same as that employed while defaecating and the urine is expelled more forcefully by the female than the male. The amount of urine passed is usually between 10 and 15 kg for the 24-hour period.

TERRITORIALISM

Cattle show territorial aggressive acts by butting or threatened butts. In defending their head space they 'hook' with sharp lateral head swings. Bulls are prone to practise more spectacular territorial behaviour than cows. The territorial display of the bull is a 'fixed action pattern'. In this they dig the soil with their fore feet, scooping loose soil over their backs. They horn ruts into the soil, rubbing their heads along the ground. At the conclusion of the display they stand and bellow repeatedly. It is common for a specially selected site, or stand, to be used for this display which may be a prominence of land.

Beef cattle under high crowding conditions in pens show clear preferences for certain locations within their pens. Such animals tend to position around the outer edges and corners of pens. The central areas of pens are less often occupied. The fact that the animals position to the outside suggests that the ratio of perimeter to area of enclosure is important. The more crowded the animals, the greater the importance this relationship is likely to be in alleviating crowding. The amount of area needed per animal also depends on features of the enclosure. Beef animals on slatted floors require less area per individual than others on solid floors. In a sense, the area beneath the slats serves as extra space, for the needs of evacuation. Increasing the viewing area, across aisles for example, serves to meet some spatial need.

REST

Cattle chewing the cud usually rest on their sternums, a posture which facilitates rumination by increasing abdominal pressure. They rest for about nine to twelve hours of the 24-hour period, sometimes idling, sometimes lying. In lying they often show an individual preference for one side rather than the other. The fore limbs are curled under the body and one hind leg is tucked forward underneath the body taking the bulk of weight on a triangular area enclosed by the pin bone above, the stifle joint and the hock joint below. The other hind limb is stretched out to the side of the body with the stifle and hock joints partially flexed. They will occasionally lie with one or other fore leg stretched out in full extension for a short period. Cattle will also occasionally lie fully on their sides, but do so only for very short periods while holding their heads forward, presumably to facilitate regurgitation and expulsion of gases from the rumen. Adult cattle take up the sleeping position seen in calves with their heads resting inwards to their flanks. This is certainly a normal resting and sleeping position taken by adults periodically, although it is also a posture typically adopted in milk fever.

Although they 'idle' in stationary standing poses, cattle, unlike horses, are unable to rest satisfactorily in upright stance for extended periods. They therefore become fatigued more severely when movement or husbandry disturbances inhibit recumbency. A minimum sleep ration of about three hours appears to be necessary, on a daily basis, for the preservation of health.

21　Behaviour Patterns in Sheep

The day-to-day pattern of feeding, rumination and allelomimetic behaviour in sheep depends largely for its variability on seasonal factors, breed of sheep, geographical situation of available land and the nutritional and chemical qualities of the grazing pasture.

REACTIVITY

The general behaviour patterns of sheep which have been described are aspects of behaviour which occur, commonly, from day to day and are not affected by season changes. There are, however, several important types of behaviour which occur during a particular season, often associated with breeding or sexual behaviour. During the summer, if it is very hot, sheep may graze more at night than during the day. Conversely, in winter they do more grazing in the hours of daylight. In autumn there is a large increase in sexual behaviour and consequently fighting or agonistic behaviour between the rams. Rams engage in running towards one another and butting; they may buffet each other with their shoulders, engage in bunching and running activities or emit snorting sounds and paw the earth with their forefeet. This may partly explain the movement of rams from one restricted area to another, which occurs in autumn, and also the unsettled nature of ram groupings with regard to the members of the group.

Vocalization

In the mating season, rams emit a hoarse 'baaing' sound when approaching ewes. Vocalization also plays an important part in the maintenance of communication between members of the flock. If a mother is separated from her young she will 'baa' until they are brought together and the young do the same. Even adult members of the flock, which become disengaged from the others, 'baa' and with increased vocalization become more animated in attempts to locate the main flock. Increased vocalization has been shown to be accompanied by increased mobility. However, while in a young lamb separated from its dam, or an adult sheep separated from the main flock, vocalization is fairly intense initially, it declines after about four hours of continuous separation.

INGESTION

The general features of ingestive behaviour of sheep are those common to all farm animals. There are periods of movement and eating as well as of drinking, idling and lying down and ruminating, interspersed with periods of intensive ingestive activity. Some animals eat less than others, others spend more time lying down, but there are particular features of behaviour which typify the sheep.

Grazing activity is largely confined to the daytime, and the onset of grazing is closely correlated with sunrise. Grazing is concentrated during the whole daylight time available, but sheep do not graze continuously. They have specific stages, during the 24-hour daily cycle, when ingestive intake is very high and others when grazing is punctuated by ruminating, resting and idling (Fig. 70). The ruminating behaviour of sheep varies from breed to breed. Some breeds prefer to remain in a particular part of the available grazing space where the nutritional quality of the grass and plants may be very high. Other breeds prefer to split up into groups throughout the area, occupying particular spots. Breeds in which ruminating behaviour is more marked, and which prefer to move about the available space, may find that some parts of the land do not provide appropriate herbage. These animals continue to move about, even though some of the available food may not be to their liking.

The longest and also the most intensive periods of grazing take place in the early morning and from late afternoon to dusk. The number of grazing periods over each 24-hour cycle averages four to seven and the total grazing time usually amounts to about ten hours. Although adult sheep usually eat more than do lambs, the pattern of their ingestive intake is less uniform and in fact some may eat less than the lambs, while others consume a comparatively large quantity of food and water. The number of rumination periods may amount to 15 during the 24-hour cycle. Although the total time of rumination may be from eight to ten hours, the length of each period may differ vastly: from one minute to anything up to two hours. The adult intake of water is from three to six

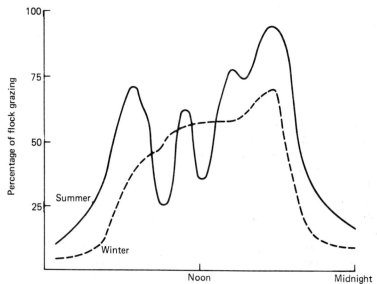

Fig. 70. Typical daily patterns of grazing activity in sheep under summer and winter conditions. The summer grazing pattern shows more intensity at dawn and more variability throughout the central span of the day. Both have a peak of activity at dusk.

litres and the number of urinations and defaecations total approximately nine to thirteen and six to eight respectively. The average grazing intake of an individual sheep may differ greatly from that of the main flock and the amount ingested may also be affected by the presence of lambs. It is widely recognized that sheep prefer certain plants and grasses to others and, of the plants that they prefer, some may prove more palatable to the animal than others. The fertility of the soil, the use or non-use of fertilizers, the geographical situation of the grazing land and the nature of the climate, all affect the grazing behaviour of the flock.

In cases of restriction sheep nibble the grass very close to the ground and faeces are often deposited where there is good quality grazing. Sheep do not normally consume plants or grass which have been contaminated by faeces, but when circumstances dictate, the sheep do consume the good herbage in spite of contamination.

The feeding and drinking habits of sheep are in some ways affected by ambient temperature, quality of grain and specific breed. It is known that, as the temperature drops, a sheep's food intake correspondingly increases; but if it becomes very cold the animal's appetite becomes inhibited and the intake declines. Sheep are known to have developed senses of smell, taste and visual recognition of food, but their intake is almost unaffected when they are made to eat without the aid of vision. They do not often, however, eat plants which are hairy or greasy.

It has been suggested that sheep are able to select what items of food they eat and thereby correct any nutritional deficiencies or excesses. Certainly, sheep with definite nutritional deficiencies have been known to correct their nutritional balance by consuming plants and grains which will do so. Sheep generally accept food of a balanced quality rather than food of a more erratic composition. Findings in this field have not been uniform, however, and many sheep are found to persist in eating a particular crop or plant which further upsets an already unbalanced nutritional condition.

Nevertheless, one evident aspect of the feeding activities of sheep is that they are fairly adaptable regarding the plants, grass and crops which are made available to them. They do tend to develop particular likings for certain crops and may even prefer one type of food, which is exactly the same in content as another but prepared differently. They are, however, also able to adapt themselves to a particular species of grass or plant if there are no other nutrients to be found and, eventually, the disliked food plant apparently becomes palatable. The stage in its development and its body weight have an effect on a sheep's food intake, more so than adjustments to sensory faculties. On average sheep consume food equivalent to 2·5% of their body weight per day.

Possibly as a result of allelomimetic behaviour, sheep often form specific paths to water sources and follow a recognized route rather than a direct one across grazing land, regardless of the time factor involved. They also generally tend to frequent a particular water place. As with feeding, the amount of water consumed varies according to breed, quality of pasture and weather conditions, for seasonal changes greatly influence the amount of water consumed by drinking.

Rumination

Cud-chewing periods number about eight per 24 hours and in this respect are not very different from the ruminating periods of cattle. Rumination in sheep, however, occurs at irregular intervals throughout the night and day and, although there may be a higher frequency of rumination early in the morning and fairly regular rumination in mid-afternoon, these tendencies are not marked and cannot be said to be characteristic of all breeds of sheep. It has not yet been ascertained exactly what induces the onset and cessation of rumination, but neural stimuli would seem to be involved. The consumption of chopped hay apparently invokes more frequent rumination than long hay and sheep, when fed small quantities of food at regular intervals instead of receiving one large feed, show a marked increase in rumination and regurgitation.

Suckling

One of the most noticeable behavioural aspects of a sheep in relation to newborn is the strong maternal relationship which often develops. A dam nearly always vigorously rejects any attempts by other lambs to engage in sucking and will look after her own young exclusively.

Most newborn lambs are able to stand within the first half-hour following birth and nearly all are able to stand within the first two hours. The lamb's first attempts to suck are usually unsuccessful; it often seeks out the teat by nosing between the forelegs of the dam or any nearby object which the lamb may feel has maternal properties. If, at this point, the newborn fails to find the teat or is prevented from doing so by the behaviour of the mother, its drive to suck may become inhibited.

Within about one hour of birth approximately 60% of newborn lambs have begun to suck and, in normal cases, nearly all lambs have sought out the udder within the first two hours. Once the newborn lamb is able to stand, it sucks and nibbles at any object which is at hand; this is usually the coat of the dam. While the dam is removing the placenta from the lamb, the latter finds its way to the region of the teats and udder. Sometimes the lamb is prevented from undertaking suckling by the diligent efforts of the mother to remove the placenta. If the udder is too large, the newborn may find the teat difficult to grasp. However, once the newborn can facilitate milk let-down by pushing the teat upwards into the udder with its mouth, progress in locating the teat again and sucking becomes very rapid. In the first week following birth, lambs suck very frequently, sometimes on sixty to seventy occasions during the 24-hour period. The duration of suckling at this time is usually from one to three minutes, but later on the young are seldom allowed by the dam to suck for periods of over 20 seconds.

Single lambs do not usually favour one teat over another. In the case of twins though, each lamb does develop a preference for one particular

Fig. 71. Nursing ewe with lamb. Note the direction of the lamb's hindquarters towards the ewe's head and the tail-flourishing by the lamb.

teat, but this may change where the other twin objects. In cases where one of the twins is removed after a period, the remaining lamb begins to suck from the other teat too. Sometimes the dam facilitates suckling by lifting her hind leg on the side at which the newborn is attempting to suck. One aspect of behaviour characteristic of the newborn lamb is the vigorous wagging of its tail when engaged in suckling (Fig. 71). It has been postulated that this is a mechanism which entices the dam to smell the anal region and so recognize her young, for some dams do not discourage the approach of another newborn if it is similar in appearance to her own.

EXPLORATION

In temperate conditions, a flock will set out into the pasture, moving together and then fanning out. However, although they may move some distance from each other, sheep often form subgroups within the main flock and continue to exist as a concerted group, following a regular pattern of movement around the grazing land.

KINESIS

The distance travelled by sheep while grazing is affected both by the immediate environment of the flock and by genetic differences affecting the behaviour of different breeds in adapting to the particular habitats in which they were reared. Thus Cheviot sheep travel further than Romney Marsh when both subspecies are kept in hilly country but only a little further when both graze on flat areas. Ewes of Hampshire breed travel less than those of Columbia or Rombouillet breeds. The distance covered by sheep, under temperate conditions, usually remains constant. The only major alteration to this behaviour occurs shortly before the breeding season when movement, particularly amongst the rams, becomes less regulated. Generally sheep travel about 8 to 16 km per day, although any increase in the duration of grazing time results in a corresponding increase in distance travelled. Likewise, any increase in available feeding space has a similar effect. These effects, however, are usually temporary and cause no major alterations to the overall behaviour pattern of the flock.

ASSOCIATION

An important feature of behaviour in sheep is their marked allelo-mimetic activity and well developed practice of social coexistence. In a comparative study of two separate flocks of sheep kept on the same grazing pasture for several days and nights, there was found to be no difference in the start or finish of the major grazing periods. Likewise, it was noticeable that no single animal was the instigator of any grazing or resting period. Although individual sheep rested or grazed at different times from the main flocks, the presence of any dominating animal which influenced the pattern of behaviour of the combined flocks was not noted.

BODY CARE

There appears to be no recognizable pattern of defaecation in sheep. Urination often occurs when animals are disturbed.

TERRITORIALISM

In the main, sheep tend to form groups which remain in a particular area. In hilly country sheep 'bed down' on the hills themselves and move to the hillier areas, if possible, during winter. In hot weather they seek out the shade of bushes and trees and areas close to water. Although these home ranges are specific in the areas covered, neighbouring sheep may share common ground and small numbers of sheep often settle down along a boundary, or in a corner of a small grazing area common to another small group from an adjacent pasture. Although rams also have specific areas in which they tend to remain, their boundaries are less clearly defined than those of ewes, and they may change their locale just prior to the breeding season.

REST

Sheep are awake for 16 hours per day. They drowse for about four and a half hours per day, far less than cattle. SWS occupies three and a half hours per day and REM sleep occurs in seven periods for a total of 43 minutes.

22 Behaviour Patterns in Pigs

REACTIVITY

The consequences of agonistic behaviour between pigs are most severe among the adults. If a strange sow is introduced to an established group of sows, the collective aggressive behaviour of the group directed at the stranger is likely to be so severe that physical injury may be caused which can even result in the death of some animals. When two strange boars are first put together, they circle around and smell each other and in some cases may paw the ground. Deep-throated barking grunts may be made and jaw snapping engaged in, at this point. When fighting starts, the opponents adopt a shoulder-to-shoulder position, applying side pressure against each other. Boars, when fighting each other, tend to use the side of the mouth, permitting the tusks to be brought into play as weapons. In this fashion, fighting boars attack the sides of their opponents' bodies. Fighting may continue in this form for up to one hour before the submission of one animal or the other. The loser then disengages from the conflict and turns and runs away squealing loudly. With the other's dominance established the encounter ends. Mixing adult pigs together, therefore, is an operation which must be carried out with care. Tranquillizers and aerosol masking odours are used in commercial pig breeding and other pig establishments to minimize agonistic behaviour among pigs when mixing them.

INGESTION

The ingestive behaviour of pigs is evidently closely linked with their exploratory drive. Out of doors the exploration of territory largely involves rooting activities. Rooting is certainly the salient feature of ingestive behaviour in pigs. Even when pigs are fed with finely ground foodstuffs they continue to show rooting activities. The snout of the pig is a highly developed sense organ and olfaction plays a large part in the determination of behaviour, not least of all in feeding activities. Pigs are omnivorous and, at free range, eat a variety of vegetable materials. They may also eat some animals such as earthworms. Under modern systems of husbandry, however, it is usual for pigs to be fed on compounded feedstuffs. Pigs consume a sufficient quantity of food of this type for 24 hours, in 20 minutes of each day. When provided with this food in feed-hoppers the time spent each day on feeding may be somewhat longer.

The quantity of food that pigs consume is marginally affected by the palatability of the feedstuff. They appear to prefer feedstuffs with some sugar content. Preference is also shown for other constituents such as fishmeal, yeast, wheat and soya bean. Substances which reduce the intake of food include salt, fat, meatmeal and cellulose. As a general rule, pigs appear to eat wet foodstuffs more readily than dry ones, though much depends upon palatability. Under management conditions where pigs are hand-fed they typically show hunger when feeding time approaches and it is evident that the temporal arrangement of their feeding activities is very well defined. Community feeding has various effects on behaviour; feeding behaviour is evidently stimulated by the sight of other pigs feeding. Pigs in groups are found to consume more food than pigs kept individually. Well-grown animals kept in pens in groups of six to eight should, therefore, be given enough feeding trough space to consume their own ration of food, without adjacent competitors being able to poach off them. It is estimated that pigs of approximately 90 kg live weight should have a minimum trough space of 35 cm each. For pig groups of this size, fed by a system of self-feeding hoppers, it has been estimated that one self-feeder is required for every five animals. Even when the hoppers are well filled, if these are too few, the pigs will not be able to avoid competition in obtaining their full daily quota of food.

Appetite can have a genetic basis; this is certainly true of pigs and some breeds have greater hunger drives than others. It has also been noted that certain families of pigs have stronger feeding drives than others and that pigs from these families are usually faster growing. The selection of highly productive pigs, in many cases, involves very little more than selecting those genes which are basic to the feeding drive.

Feeding drives are often keen amongst breeding sows. These animals

are regaining body weight which was lost during the preceding lactation and, for this reason, there is more competition amongst them than others during feeding time. The introduction of individual feeding stalls for sows has been a great help in dealing with this situation. The stalls operate best when sows are allowed access to a communal exercising area between feeds.

Self-feeding pigs randomly space their eating and drinking periods throughout the day. Of the two, eating is the preferred activity. Pigs quickly learn to drink from mechanical devices which supply water when some plate or button is pressed. Water drinking is influenced by both animal size and environmental conditions. Under normal conditions of management, full grown pigs consume approximately 8 kg water daily. Pregnant sows may drink more water than this, their consumption usually being in excess of 9 kg per day.

EXPLORATION

Much of the general activity of pigs appears to stem from their exploratory behaviour. They show a very well developed exploratory drive, most of which is directed at objects at floor level which are investigated by smelling, nibbling and rooting. Such actions may have a destructive effect on objects, where these are subject to continuous investigation by a group of pigs, which are severely restricted in their movements. This is apparently the underlying cause of the vice of tail-biting. The tendency to over-investigate specific items in their environment can be controlled to a degree by providing additional objects for investigation and some pig breeders, aware of this fact, provide growing pigs with objects to be used as toys. These include motor tyres and chains suspended from the ceiling. The provision of straw bedding also provides an alternative outlet for the investigatory activities of pigs.

KINESIS

While young pigs are nimble and able to run about, the mature pig, with its relatively massive trunk, is physically ill-equipped for movement at speed. Well grown pigs therefore run distances of only a few metres. They are, however, able to trot at a reasonable speed over long distances.

Pigs have long periods of inactivity each day. During these periods they typically rest in huddled groups. Various studies have shown that pigs are active during the night hours. This appears to be so particularly in the case of sows in estrus.

ASSOCIATION

The social organization of groups of pigs is known to depend upon the establishment of a social hierarchy. For the social hierarchy to function

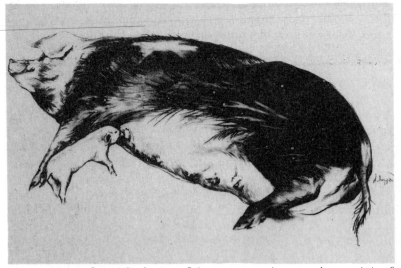

Fig. 72. The preferential selection of the upper anterior teat: characteristic of 'teat ordering' in nursing behaviour of pigs.

properly, the size of a group and the space allocated to it are important. It is also necessary for the members of the group to be capable of prompt recognition of each other. In pigs, it is still uncertain how the mechanics of recognition operate, though it is evident that different types of recognition exist. A form of face-to-face recognition appears to operate during an initial introductory period in the formation of a hierarchy. Sensory clues such as olfactory stimuli are probably involved in the maintenance of the social hierarchy. It is also evident that pigs in an established hierarchial arrangement are quickly able to recognize an alien in the group. Visual and olfactory cues seem to be the principal differentiating features of pigs for each other.

Since pigs are normally kept in groups which are allocated a relatively small area it is inevitable that many social behavioural exchanges occur and that many of these have aggressive outcomes. The role of aggression in pig production is of considerable importance. By about two weeks of age, piglets begin to show exchanges of agonistic behaviour in the form of brief but vigorous fighting. As a consequence of this, dominance–subordination relationships are quickly formed within the litter. New relationships of this kind form each time strange pigs come together.

Fig. 73. An established teat order in piglets three days old. One piglet in an inferior, posterior position shows facial wounds from prior fighting, evidently unsuccessful.

Where there is no mixing of this kind, social organizations become well established and functional and they reduce aggression within the group. Limited aggression within a group of pigs conserves energy. This in turn reflects in production. Dominance in the hierarchy ensures that an individual is able to carry out chosen behaviour. When competition for food or space increases, aggressive behaviour within the group increases. The build-up of aggressive behaviour leads to a weakening of the social hierarchy and, in due course, may cause it to break down completely. Some forms of abnormal behaviour in pigs are likely to be associated with pathological increases in aggressive behaviour resulting from instability in the social hierarchy.

BODY CARE

Through their physical characteristics pigs, more so than any other farm animals, are susceptible to heat stress. In the presence of a high ambient temperature, they are singularly ill-equipped physically to radiate heat. Given access to water, they engage in wallowing activities (Fig. 74) during which they wet all the ventral areas of the body and give up excessive body heat by conduction. Under field conditions in warm

Fig. 74. Wallowing behaviour in a sow, outdoors in a tropical environment.

weather pigs create mud-wallows. Wallowing in mud, in addition to permitting heat to be conducted from the body, also causes a reduction in body temperature by radiation; this allows pigs to forage extensively in excessive ambient temperatures. Studies on the development of heat stress in pigs show that they can develop hyperthermia within a very short space of time; but, providing that it is possible for the body surface to be wetted, hyperthermia can be controlled to the point where animal health and comfort are not impaired. Close studies have been made on the effect of wallowing on a rapidly developing state of hyperthermia in sows in the tropics. In some instances it was noted that actual heat stress was brought under control by the pigs' voluntary wallowing activities.

Because of their morphological characteristics pigs are ill-suited to extremes of ambient temperature. They receive poor body insulation from their sparse covering of hair. They have very little loose skin from which to radiate excessive body heat and their sweat glands are confined to their snouts. In spite of these physical inadequacies, pigs are able to adjust to temperature extremes through specific thermoregulatory behaviour.

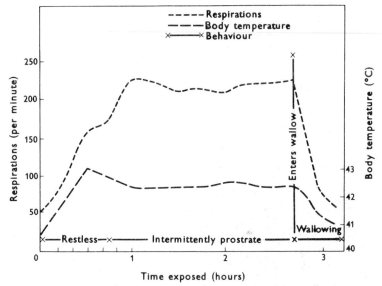

Fig. 75. Behaviour in relation to incipient heat stress in a pig. Wallowing is seen to control the syndrome.

The ability of newborn piglets to adapt to their environmental temperature is very limited as they are prone to lose body heat rapidly. The behavioural mechanism for dealing with this problem is huddling. From the time of birth, young piglets display huddling behaviour as an organized attitude for most of the day. During huddling they lie parallel to each other, often with head and tail ends alternating along the row. Whilst lying closely together side by side, they usually have their limbs tucked underneath them. When a group is large, some of the piglets in the middle may overlie others. The result of this huddling behaviour is that the quantity of heat lost by the piglets is much less than would be lost otherwise. Although huddling behaviour is characteristically shown in the litter early in life, it is nevertheless a behavioural pattern which is retained by groups of piglets, into adult life, as a means of conserving body heat.

Elimination

Elimination does not take place at random in the pen; specific sites are chosen by pigs for defaecation and urination. In spite of a reputation to the contrary, pigs are extremely clean in their habits if the system of husbandry imposed upon them gives them the opportunity to express their normal behaviour patterns. Pig premises which are appropriately designed to create dunging areas are usually properly used by pigs. Pigs apparently have a keen sense of territory and, even in the most limited quarters, they reserve an area for sleeping accommodation and an area for excretion. This sleeping area is kept as clean and dry as is possible. Under conditions of crowding, it is sometimes difficult for groups of pigs to maintain organized eliminative behaviour. When growing pigs are allocated less than 1 m² of floor area each, their eliminative behaviour becomes disorganized and uncontrolled. Much of a pig's eliminative behaviour is learned during infancy from mature animals and if the behaviour patterns are not acquired by learning at an early age they may not be acquired at all. Such pigs, in their turn, are unable to pass on learned behaviour of this type. These pigs are usually observed to be contaminated with their own excreta and their presence in large proportions in a pig herd is a reflection on the system of management.

Urination, apart from its eliminative function, is also useful in other contexts. When penned pigs are exposed to high ambient temperatures and their normal behavioural methods of controlling hyperthermia cannot operate, it is common for them to urinate in a part of the pen and thereafter to wallow in this urine. The passing of small quantities of urine is observed in both sexes during the precoital period.

The patterns of eliminative behaviour whilst being learned from each other are also influenced by the location of food and water sources. It is commonly found that pigs excrete close to the source of their drinking water.

TERRITORIALISM

The stocking density in groups of pigs is known to have various effects upon their behaviour. Social encounters in penned pigs normally take place at or near the feeding trough. These social encounters lead to clear-cut results when the hierarchical system has previously been well established. When growing pigs are allocated only about 0·75 m² of pen each, there is a rise in severity of social encounters. When the stocking density is any heavier than this it is found that individual pigs, which are low in the social hierarchy, are unable to avoid the consequences of aggressive encounters. In consequence these pigs suffer more injury and reduced feeding opportunities. The productivity of the unit is thus adversely affected.

REST

Of all farm animals, pigs spend most time resting and sleeping. Sleeping in groups simultaneously is usual and they are recumbent in rest or sleep for 19 hours each day. Pigs also drowse and about five hours are spent in this state daily. Of the total sleep time, SWS occupies six hours per day and REM one and three-quarter hours in 33 periods.

Pigs are characterized by extreme muscle relaxation during sleep. While this is difficult to appraise in a mature sow, young piglets exhibit profound relaxation. When a sleeping piglet is picked up, it is found to be totally relaxed and evidently in a deep state of sleep.

23 Behaviour Patterns in Poultry

REACTIVITY

Displays are common in the behaviour of the domestic fowl and these displays are often accompanied by particular vocalizations. Seven recognizable forms of display occur in reactive behaviour in the fowl. These include: feather-ruffling, in which the neck is outstretched, feathers are ruffled and the whole body is shaken; headshaking, in which the head is tilted to one side and shaken vigorously in a circular fashion; waltzing, in which one wing is dropped as the cock bird advances towards another bird in a sideways manner; wing-flapping, in which the bird stretches to its full height and flaps its wings repeatedly; tail-wagging, in which the tail is extended behind the bird and wagged from side to side; circling, in which one bird circles another with high-stepping action; and tidbitting, in which the bird pecks at the ground, scratching it intermittently.

In agonistic activities a variety of behaviours and calls occur. Two types of display are shown after fighting (Fig. 76). In one form the victor adopts one of several postures which have the common characteristic of having high tails with open tail feathers. In the other form the loser of a contest adopts one of two postures in which the tail is held level or down with the tail feathers closed. Among agonistic activities, features such as circling, pecking, leaping, hopping and fighting are common. Oblique and side postures are often shown towards antagonists. In association with fighting a variety of 'irrelevant' activities may be shown such as head-shaking, ground-pecking, preening, head-zigzagging and strutting.

Vocalization in the domestic fowl has been recognized to be quite complex. Some observers have reported 33 types of vocalization ranging

Fig. 76. Postures in poultry following agonistic encounters. The postures in the upper row, with raised tails, indicate a winning outcome, while those in the lower row, with closed tail feathers, indicate a loss.

from the most common one of crowing to special calls by broody hens and laying calls.

INGESTION

In addition to pecking and swallowing minor variations occur in the ingestive behaviour of the fowl. Free-range poultry, when they grasp a large food object in the bill, may run with it while 'peeping'. Again on free range the domestic hen typically makes two or three backward scratching movements with alternate feet before stepping back one pace to peck at the area of ground which has been disturbed. Poultry typically peck at their food with jerky head movements directed like small hammer blows. Sometimes poultry will sweep their bills in hoppers of soft feed, displacing it from side to side. In the typical food-pecking action the bird's eye becomes closed at the time of the strike. The pecked item such as grain is then grasped between the mandibles and, following this prehension, the head is jerked upwards and backwards as the food is swallowed. The appetite of poultry has been subject to enormous study and is evidently complex. Birds do experience hunger behaviour after fasting as shown in pecking rates, for example. Undoubtedly young chicks in association with their hens learn certain items of feeding behaviour from them. In the selection of food, visual stimulation evidently plays a part and food preferences are recognized. Certain cereal grains such as wheat are apparently preferred to others such as oats. Poultry have specific hungers and studies on the fowl indicate that this specificity is a means whereby feeding can be adjusted to special bodily needs. A need for calcium is a notable one, particularly in laying birds.

Poultry drink frequently each day and some studies have shown that fowl will visit a drinking fountain in their pen thirty to forty times daily.

EXPLORATION

The domestic fowl in most modern systems of husbandry has no opportunity for exploratory activities, but birds at free range show a considerable amount of exploratory behaviour in the form of food-searching which may take them several hundred metres away from their home site in the course of each day in extreme cases. Birds also seek out suitable ground in which to settle and dust bathe. Visual inspection of strange objects is carefully practised by the fowl. The fowl's vision is evidently good and it has been determined that the hen's colour vision is trichromatic and similar to human vision in the appraisal of colour. Under natural conditions hens will often seek out nesting sites in locations quite remote from the core area of their home range following territorial exploration.

KINESIS

Locomotion in the hen is by walking and running, but many hens are capable of flying distances of about 10 to 15 m. Much kinetic activity is invested in pecking for food but exercise activities are also common. These include vigorous extension of one wing after another. More usually when one wing is outstretched in a backward direction the leg on the same side may also be extended backwards. Wing flaps represent another form of exercise. Some forms of caging in current use virtually prevent these activities.

ASSOCIATION

The early associative behaviour of the chick has been described in the section dealing with behavioural development. The most notable form of associative behaviour in the fowl is the peck order, which has been well observed in studies over the past half century. In the fowl the social dominance order takes the form of a 'peck right' type of hierarchy. In its simplest form this social order, maintained by aggressive pecking of each other, is shown by one hen dominating all others in the flock, the second in status dominating all but that one, the third in status dominating all but the two above, and so on. This produces a linear hierarchy. In many flocks the order is not linear throughout and triangular situations of dominance may occur in which a bird in a subordinate position may find itself able to peck with impunity a bird higher in the order than the one immediately above. In other situations the departure from a linear system may be even more elaborate. It has long been recognized that in some peck orders there may be despots creating some socially suppressed individuals at the bottom. Birds low in the peck order generally attack their inferiors more frequently than do individuals which are high in the order. It is also noted within the peck order that great contrasts exist in the amount of tolerance existing between pairs of birds. When both sexes are present in the flock a separate peck order becomes established for each sex group.

A notable feature of the peck order is its determination of feeding priority when there is competition for food. Subordinate hens in a flock may not have access to adequate food and produce less if feeders, waterers and nesting boxes are not adequate in number and distribution to ensure that they have full opportunity to feed and lay. When the peck order in a flock is subject to considerable disorganization, due to removals and additions of birds, the entire flock consumes less food and is a much less productive unit, with many individuals growing poorly and losing weight. Hens in unstable peck orders lay less. In large flocks separate peck orders may become formed in subgroups. A single peck order is unlikely to include more than 40 birds. Strange birds introduced to an established flock are initially subject to considerable aggression and are only slowly introduced to a low level in the peck order. Their position in the order can be increased only with aggressive fighting.

Debeaking is commonly employed to minimize pecking but it does not eliminate aggressive pecking behaviour or prevent the development of a peck order. Pecking among debeaked birds is, of course, less traumatic and does not disturb subordinate individuals excessively. Crowding hens on wire mesh floors often prevents a peck order being established due to reduced mobility of birds, but feather-picking is common in such circumstances.

In domestic turkeys, social organization is based on flocking tendencies. Flock groups are organized according to a social dominance order but this is less stable than that of the fowl. In groups of mixed sex males dominate females. In male turkeys penned together, however, changes in rank may occur every few days. Certain breeds or varieties of turkey tend to be dominant over others when mixed. Bronze turkeys are normally dominant over grays, for example. In the agonistic behaviours which determine dominance, the most common is a simple threat, with one bird promptly submitting to the other. In more balanced contests, both birds may warily circle each other with wing feathers out-spread, tails fanned, each emitting a high-pitched trill periodically. In fighting, turkeys leap into the air and attempt to claw each other. The winner is usually the bird which can push, pull or press down the head of the other. Tugging battles may develop and, since richly vascularized skin areas may be torn, blood may be shed. Physical injury is usually slight, however, and birds do not fight to the death. Low-ranking birds which do become wounded must be separated from the group since other turkeys will peck and aggravate their injuries.

BODY CARE

The fowl's principal activities in relation to body care are concerned with comfort-seeking, evacuation and grooming. Among comfort-seeking activities are various acts of stretching and flapping wings outward, forward and downward and extending limbs. In hot conditions hens will sometimes sit with their wings opened out from the body and drooping; this is evidently for thermal control.

In evacuation, the cloacal anatomy of the bird dictates only one form of excrement. In dropping this, the bird adopts a sitting posture briefly and expresses a single deposit. While this is done in an apparently random way, as regards time and location, much evacuation is done during roosting.

Grooming is typically in the form of preening and this too is apparently practised whilst roosting, more than at other times. Preening is a significant activity, even in parasite-free birds, in which it may occupy a total time of about one hour each day. In preening the bird disturbs its feathers with its beak and, with short brisk movements of the head, picks or scrapes at localized sites of skin. The neck, breast, limbs, back and wings receive this attention. Dust-bathing is a grooming behaviour in which the bird, after making a shallow hollow in the ground, sits in it and rubs the lower parts of its body into it with some vigour. If the hollow has loose material this becomes agitated and worked into the ventral feathers. Sometimes in dust-bathing a bird will lie to one side and work the body into the shallow and may also rub the head and neck into the ground. Dust-bathing is subject to social facilitation and a dust-bathing bird is often joined by others to engage in a communal session of this body care activity. Following a dust-bathing session the bird normally engages in general body shaking which dislodges much of the dust picked up. Other forms of shaking include feather-ruffling and tail-shaking. These shaking actions may occur independently of dust-bathing. Hens in battery cages show essentially similar behaviour, including 'dust-bathing'.

TERRITORIALISM

Cock crowing is considered to be a territorial pronouncement. With adequate territory cock crowing is usually limited, notably to early morning. Where cockerels are crowded together and actual territory is nil, as in battery systems for artificial insemination purposes, excessive crowing is a remarkable and disturbing feature. Large flocks of hens do not have an even dispersal of their individuals, their spatial distribution is evidently not random. A territorial arrangement of flock members seems dependent on facial alignment of nearby birds so that a pre-ferred angle of between 90° and 180° is kept. Birds which may be generating additional stimuli through physical or behavioural characteristics are given greater distance. Territorial organization of flocks is evident in free-ranging birds and, when territories overlap, agonistic encounters will occur in that area. It has been found that hens in laying cages, given extra viewing area, show less behavioural evidence of stress.

REST

Drowsing occurs in occasional daily periods in poultry. The hen drowses in a squatting position and drowsing phases are interspersed with true sleep when birds perch on roosting places at night. Poultry resting and sleeping on perches draw the head and neck close to the body and grasp the perch firmly with their feet, maintaining this position for several hours. In roosting, preference is shown for locations high from the floor area but such perching is not possible in a battery cage system. In cages with sloping floors there is a preference to sleep on the highest available place on the slope. Though this is apparently a poor substitute for a perch pole, fowls evidently adapt to such facilities, probably at some physical expense. The fowl tends to go to roost about the time of commencement of evening twilight and continues to roost till the termination of morning twilight. Sleep does not occupy all of this time and most birds continue to roost after waking. Sleep is evidently determined by a combination of light and other environmental stimuli, photoperiodic factors and an internal biological rhythm meeting the physiological needs of sleep.

Part V

Reproductive Behaviour

24 *Sexual Behaviour*

MALE

Male Sex Drive

Sex drive in the entire male amongst farm animals is commonly referred to as libido. This drive develops at puberty and, after maturation, persists at a fairly constant level for the remainder of the animal's life-span. Libido or male sex drive is dependent, basically, on the production of testosterone by the testis, but in an individual animal the typical level of the sex drive is predetermined by inherited characteristics.

Some variations in the level of male sex drive are to be observed and these can have quite considerable consequences in farm economics. In the bull, libido varies in degree between age groups and between breed types. In general, a lower level of libido is found in beef than in dairy breeds. Comparing the species, the highest levels of libido are generally noted among the seasonal breeding animals, for example rams. Clearly those species which concentrate their breeding season into relatively short periods require high levels of libido for effective reproduction during that time. The level of sex drive may change as a consequence of various factors: there are physical changes that occur in ageing bulls which are known to reduce their sex drive. Quite clearly, an animal which is experiencing discomfort or even pain during movement in mounting will, in time, have his breeding behaviour impaired. It is also strongly suspected that adverse experiential factors can cause sexual inhibitions in all stud animals.

While low libido in free-living animals is species self-limiting, in domestication unwise selection can permit its propagation; and there is evidence that this has occurred in some of the beef breeds of cattle. It is recognized by practical flockmasters that, in some breeds of sheep, the rams have higher levels of sex drive than in others. There is also growing suspicion that certain breeds of pigs produce boars with inferior libido. Impaired or inferior libido is not always inherited. Obesity in stud animals often contributes to low libido. Some skeletal defects such as arthritis are also a common cause of poor breeding behaviour. Clinical studies in recent years have confirmed that many impotent boars, for example, suffer arthritis of the hip joint.

An interesting phenomenon of male sex drive is one which is associated with its total absence. When entire male animals experience complete loss of their sex drive for some reason or other, they seek out the company of other male animals of the same species. The bachelor groupings which result when such male animals gather together are a phenomenon which has been observed in many species of free-living animals. Bachelor groupings can be noted when large numbers of bulls run together. They are also seen among rams during the long non-breeding season of the year. The purpose of bachelor groupings is to suppress further any libido that individual males might have and such close male groupings quite clearly show successful adaptation amongst their members.

Libido manifests istelf in a variety of behavioural components which comprise male courtship. A behavioural component of libido common to most ungulates is the 'olfactory reflex' or *flehmen* (Fig. 77). In this, the animal fully extends the head and neck, contracts the nostrils and raises and curls the upper lip. It occurs most usually subsequent to smelling urine and nosing the female perineum and is almost certainly a form of odour testing.

'Prompting' and 'tending' are often shown by the male while consorting with the female before and during estrus in the latter. Firm standing by the female is the positive response to 'nudging' by the male subject and provides reciprocal stimulation. Nudging can be seen in some form in the precoital behaviour of most ungulates. It is seen in the stallion, for example, which tests estrus in the mare not only by smelling her, but also by biting and nipping her over areas of the body, working from the hindquarters towards the neck. Precoital nudging is also commonly observed in the male goat.

Once mating has occurred, freely associating partners often exhibit the so-called 'tending bond'. Both sexes contribute to this temporary alliance, thereby facilitating repeated mating and ensuring optimum conditions for fertilization.

Many precoital components of behaviour tend to be species-specific. Sheep and goats, however, have items of courtship behaviour in common. These include: nosing of the female's perineum, nudging the female, *flehmen*, flicking out of the tongue, striking out with a fore limb and low-pitched bleating sounds. In addition to these behavioural

Fig. 77. *Flehmen* in male ungulates. An olfactory reflex involving the vomero-nasal organ (Jacobson's), a pair of blind-ended tubes on the floor of the nasal cavity and communicating with the nostrils and the palate. Believed now to be a specialized organ for the detection of pheromones or stimulating odours.

features, the male goat also spills small quantities of urine, particularly on to his forelegs. Butting of the female's hindquarters is also occasionally seen in both of these species. False mounting attempts are sometimes shown by rams and billy goats; this behavioural feature is also seen in horses. Bulls often pump their tail-heads up and down during precoitus and during the same period may pass small quantities of faeces.

There are, however, certain major behavioural activities common to all male farm animals. These are: threat displays, challenges, territorial

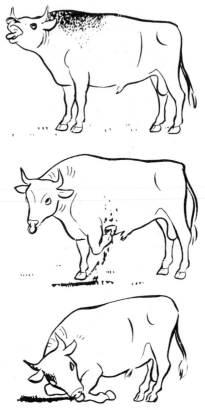

Fig. 78. Three components of the full behavioural pattern of male display in the bull.

activities, female seeking and driving and female tending. These behavioural activities tend to flow into each other.

Threat displays are usually produced by animals in a static posture. The threat display of the bull is in fact a physiological state of fight-or-flight. In this state the animal arches his neck, shows protrusion of the eyeballs and erection of hair along the back. During the threat display the bull turns his shoulder to the threatened subject (see Fig. 36).

The threat display of the stallion involves rearing on his hind legs and laying back the ears.

Threat displays are rarely shown by rams towards humans but, nevertheless, forms of threat are exhibited in the presence of other potential

Fig. 79. Precoital interaction in goats featuring (a) nosing of perineum by the male (black), (b) alignment of both and (c) sensory intake by female.

aggressors and in these circumstances the threat display usually involves vigorous stamping of a forefoot.

Challenging behaviour among male animals is typically seen where there is an opportunity for males to form pairs. When challenges are taken up the outcome of the challenge eventually determines the 'peck-order' or hierarchy. The peck-order also affects sexual status in free-breeding groups of animals. The male at the top of the peck-order may perform most of the breeding with the available females. Under domestication this type of circumstance is not usually permitted to develop. Nevertheless, breeders are not unfamiliar with 'boss bulls' and 'boss rams'. The challenging behaviour of the bull has three main components: roaring, pawing the ground and horning the ground or solid objects. The roaring of the bull is typically masculine with its bellowing nature and its broken pitch. The pawing behaviour at pasture has a follow-through foot action which effectively scoops soil upwards, throwing it over the animal's back. The bull also rubs the side of his face and his horns vigorously against the surface of the ground and this is often practised in a kneeling position and is displayed vigorously (Fig. 78).

Territorial behaviour is demonstrated far less by farm animals than by many free-living wildlife species. The latter may de-bark trees extensively and disperse secretions around their territories. Nevertheless the domesticated male animal engages in behaviour which can be described as territorial. Pawing and horning behaviour by bulls creates bare patches of earth and these patches located throughout his territory are

Fig. 80. The 'tending-bond' persisting in a pair of breeding goats.

clearly a claim to possession of a given area. Stallions also claim territory; they do this at pasture where they urinate and defaecate at selected spots. Given suitable territory they mark it in this fashion, returning from time to time to defaecate and urinate again in the same places.

Male seeking behaviour of females in estrus goes on continuously under free-breeding conditions. Nosing the perineum and the hindquarters of females is a fairly continuous male activity. The experienced male animal is capable of detecting the pro-estrus phase in the female and

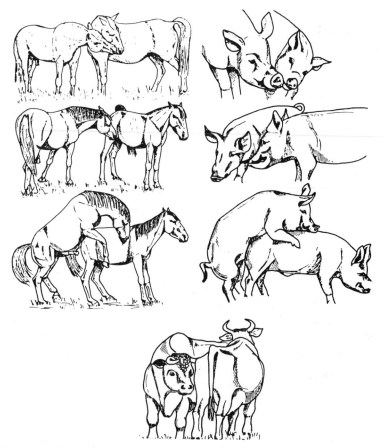

Fig. 81. Mating activities showing equine and porcine pre-coital exchanges of courtship and the 'lateral-and-opposite' stance in the bovine 'tending-bond'.

after locating a female in this state he will consort with her and will engage in 'driving' behaviour. The male may drive the female forward in differing ways. Stallions force mares to move forward by nipping their hindquarters and often by biting them in the regions of the hocks. Many male animals, such as rams and boars, will actively pursue females in pro-estrus.

'*Nudging*' in some form or other can be seen in the precoital behaviour of all the farm animals and is prominent in courtship behaviour. Nudging behaviour prompts the female to move forwards.

In estrus the female responds to this stimulation by adopting a stationary stance, so facilitating mating. Firm standing is the positive response to nudging and provides reciprocal stimulation for the male. Rams nudge by pushing with their shoulders and also striking the hind limbs of the ewes with their forefeet. Butting is another form of nudging shown by all the ruminants including bulls, rams, and goats. Boars 'root' sows.

'*Tending*' *behaviour* is displayed by farm animals when opportunities permit. The male maintains close bodily contact and association with the female whilst grazing near her. Both sexes contribute to this temporary alliance. In the tending-bonds of most farm animals there are phases when the male animal rests his chin over the hindquarters of the female. This 'chinning' behaviour is best seen in cattle but it also occurs in other species.

FEMALE

Estrus is the behavioural state when the female seeks and accepts the male. The behavioural features are synchronized with various physiological changes of the entire genital system essential for mating and fertilization. The signs of estrus are characteristic for each species but variations occur between individuals. Seasonal and diurnal variations also occur in estrous manifestations.

Behaviour in general is altered when the mating drive in the female subject is evoked. The usual routines of behaviour are disturbed during overt estrus and typically there is a reduction in ingestive and resting behaviour, while locomotor, investigative and vocal behaviour are increased. All of this is secondary to the essential character of estrus, namely receptivity to mating.

The influence of environmental factors on overt estrus is very real and was not adequately appreciated formerly. Apart from those environmental factors which ensure health, such as nutrition and

Table 6. Annual incidences of oestrus in 12 unmated ponies

Time	Range	Mean
Total days	49–103	83
Percentage of year	13·4–28·2	22·7

housing, others, of a biological nature, are also influential on estrus. Recognition of these biological factors improves the modern concept of estrus occurrence and affords improvement in the detection and management of this critically important phenomenon.

Horses

Estrous behaviour in mares shows a range of characteristics peculiar to this species. The intensity of the estrous drive varies probably more than in any of the other farm species. A mare in estrus typically adopts frequently a urinating stance. During these periods of straddling, mucoid urine is ejected in small quantities which may splash at the animal's heels. Following this, the animal maintains the straddling stance for a time with the hind limbs abducted and extended. The tail is elevated so as to be arched away from the perineum. The heels of one or other hind hoof are commonly seen to be tilted up off the ground so that only the toe of that hoof remains touching the ground. While this stance is maintained the animal shows flashing of the clitoris by repeated rhythmic eversions of the ventral commissure of the vulva. The duration of equine estrus is four to six days on average but varies considerably, some lasting only one day and others lasting up to 20 days.

Cattle

The behavioural signs of estrus in cattle include the following features. There may be an increase in what could be generally termed excitability and the estrous cow bellows more than usual. Grooming activities, in the form of licking over other animals, are also increased. Typically, the estrous cow frequently makes mounting attempts on cattle. When several cattle in a group have been prompted to mount each other, through the initial activity of the estrous cow, it may become difficult for an observer to identify the cow in the group which is in true estrus, but when one animal in particular is standing to be mounted by others it is usually the animal in estrus. In cattle estrus lasts for a period of 12 to 24 hours and it is commonly observed to be of shortest duration in younger cattle.

Sheep

The ewe shows discrete behavioural evidence of estrus. Heat in this animal is extremely difficult to detect if there is no ram with the ewe.

When a ram is present the ewe coming into estrus will usually seek out his company and consort with him for a period of many hours before true estrus commences. Many observers have noted that ewes in estrus frequently initiate the first sexual contact with rams and thereafter follow the rams about in their grazing movements as long as heat persists. Although the normal period is recognized as being just over 24 hours, estrus can last for up to three days in some ewes. Mutual riding among ewes, one of which is in estrus, has not been reported.

Goats

In goats, the signs of estrus are very marked indeed. The female in estrus shows rapid tail-wagging actions during which the upright tail quivers rapidly from side to side. This tail action resembles 'flagging' and is shown in frequent bursts which are repeated throughout estrus. Estrus lasts on average just over 24 hours. During this period the female goat eats less than usual, has a tendency to roam and bleats very frequently and loudly.

Swine

A salient feature of estrous behaviour in the sow is the adoption of an immobile stance in response to pressure on the back. In pig-breeding practice this is often supplied by the animal attendant pressing the lumbar region of the sow or sitting astride the animal. The onset and the termination are gradual and of low intensity but the 'standing' period is well defined and lasts less than a day on average. The estrous sow is sometimes restless when enclosed, this being rather more noticeable during the hours of night. Some breeds, particularly those carrying erect ears, show a conspicuous pricking of the ears in full heat. The ears are laid close to the head, turned up and backwards and held stiffly. 'Ear-pricking' is often shown when some movement is taking place behind the animal. Mutual riding is very much less common than in cattle but the subject in heat is often ridden by other females. Occasionally, among groups of sows, one particular sow will perform most of the riding.

Estrous Stimuli

It is now recognized that for estrous responses to be shown in complete form, in many farm animals, it is necessary for some form of exterostimulation to be provided. Male attendance which supplies prompting, as for example in the form of nudging, is now appreciated to be an important contributor to estrous displays in females. This male influence on estrous behaviour is termed 'biostimulation'. Although this influence has been noted by some stockbreeders for many years it was not accepted as a fact by animal scientists until the so-called Whitten effect was reported. The Whitten phenomenon is one in which estrus in mice can be synchronized and induced by the introduction of a male mouse into a colony of females. Carefully detailed experimental work on this phenomenon has established that the majority of the females come into estrus as a result of being exposed to a specific odour from the male animal. There is evidence that this type of phenomenon may also occur among the various farm animals.

It is known that the introduction of a ram to a flock of ewes can influence the commencement of the breeding season in that flock. This effect can even be achieved without the ram having physical contact with the female members of the flock. Ram influence, provided under practical conditions by 'teaser' rams, undoubtedly brings on seasonal breeding activities in ewes more rapidly than when they are left in an all-female group. The masculine stimulus may be provided by the sight, sound or odour of the ram. It would seem, however, that, whatever the stimulus, it can influence the breeding behaviour of ewes over some distance. Such odorous stimuli are termed pheromones.

The production of pheromones by farm animals has been studied most intensively in the pig. It has been found that the boar produces a substance called muskone which appears to have the effect of prompting sows to display estrous behaviour. This substance appears to be produced in a submaxillary gland but is excreted via the prepuce. Boars provide a range of stimulating factors in the form of elaborate nudging and highly specific vocalization, together with the production of pheromones, in order to induce maximum estrous responses in sows. While other examples of biostimulation in farm animals show a latency in response, this has not been noted in the pig to date. Indeed in a sow which is already in estrus, the standing response can be obtained almost immediately when some of the specific stimuli are provided, e.g. a recording of boar sounds or the dropping of small quantities of seminal fluid on the snout of the sow.

Evidence of biostimulation has also been obtained in cattle. It has been reported that, in several experiments in which teaser bulls are run with groups of newly calved cows, the so-called teased animals show signs of estrus much earlier than similar cows in control groups. Breeding behaviour may be shown four weeks earlier in the teased group than among controls. The phenomenon of biostimulation and its significance must now be more fully appreciated.

It is clear that the induction of estrus in the farm animals is due not only to internal, endogenous factors but also to external, exogenous stimulation. The latter stimulation is complex, involving odour, sound, sight and touch. Earlier it was supposed that estrus was entirely under endogenous control, but it is clear now that exogenous factors are of considerable importance. The problem of anestrus—failure to show estrus during the periods of the breeding season or life cycle when it should be shown—in farm animals might be attributed, in many cases, to the absence of exogenous factors through faulty husbandry as well as faulty observation and detection.

MATING

Mating is the consequence of the sex drives combining. The joint impetus of mating is the emergence of estrual drive in the female and the activation of male drive. The timing of mating is arranged so that spermatozoa are introduced into the female genital tract before the ovum is liberated from the ovary, a requirement of fertility. Copulations are therefore centred mainly in the earlier part of estrus, with some fewer repetitions of mating occurring in the later part. The frequency of repetition varies enormously, differences in frequency being much greater between individuals than between species and clearly dependent on sex drive intensities. As a general rule, the rate of repetition diminishes in the middle period of the day and in late estrus. Drops in sex drive help to conserve energy and sperm reserves in the male. Mating is not, therefore, a totally committed force in species survival, although it is the basis of it.

Patterns of mating behaviour of increasing complexity have evolved in mammals; these patterns are essentially concerned with stabilizing the mating relationship. The enactment of mating patterns depends on the strength of sex drive plus the strength of the stimulus from the opposite subject. Mating activities are basically instinctive, but are also partially learned. The view has been expressed that the need for learning in mating behaviour is apparently greater in the male than in the female, but it is now recognized that heifers and maiden sheep, for example, are not always efficient in converting their estrous drives into proper estrous behaviour.

Coital Behaviour

In the sexually mature male, copulating behaviour becomes orientated and directed so that the female is appropriately covered for intromission to be accomplished. This directional aspect of mounting seems to be acquired by learning, but is shown more positively by male animals which are highly stimulated sexually. Disorientation, both in the approach to mounting and in mounting itself, can be seen in male animals with a low level of libido. Male animals seldom mount the females of species other than their own but stallions will mount female donkeys and jack-asses will, likewise, mount mares. Such inter-species coitus allows mules to be bred. Occasionally sheep and goats will intermate but normal pregnancies do not result. There have been reports of isolated instances of abnormal sexual behaviour in which the male of one species mounts females of a different species. Examples include bulls mounting mares, stallions mounting heifers.

In each of these cases the animals involved were stated to have been in each other's company from early life and this suggests that the abnormal behaviour was a consequence of imprinting.

'False mounting' attempts by the male animal are commonly seen in courtship. In these instances dismounting subsequently follows quickly without any fore limb clasping or pelvic thrusting movements. False mounts show that the mechanics of mounting and of intromission are separately controlled. False mountings are to be seen in the mating patterns of the stallion, the sheep and the goat. In the stallion it is believed that some two or three false mounts are normal before effective mating is achieved.

Following normal mounting, penile intromission is effected, but this is dependent on prior penile erection. Erection in the stallion is much less rapid than in the ruminants. It may be for this reason that 'false mounts' are customarily shown by stallions before mating takes place. In the bull, goat and ram there is a more rapid erection and protrusion. In the ruminant species intromission consists only of a single pelvic thrust which is followed by dismounting. In the stallion intromission is maintained for a period of a minute or more during which there is pelvic thrusting and subsequently the adoption of a fairly static posture after which dismounting occurs.

Table 7. *Coitus in farm livestock*

	Male reaction time	Precoital behaviour of male	Manner of intromission	Duration of intromission and site of insemination	Repeat matings
Horses	Averages about five minutes	Noses genital region Genital olfactory reflex Bites croup region Penis erects fully	One to four mounts Several pelvic oscillations Terminal inactive phase	One minute Intracervical	Breeding usually arranged to permit two to four services per heat
Cows	Mode two minutes Mean 12 minutes Mean of beef breeds 20 minutes	Noses vulva Genital olfactory reflex Alignment Licks hind-quarters	Single pelvic thrust coordinated with clasp reflex	Five to ten seconds Intravaginal	Free-ranging bulls will serve cows three to ten times in heat period
Pigs	One to ten minutes	Approaches sow giving series of grunts Noses vulva vigorously Champs jaw and froths at mouth	Short protrusions of spiral penis repeated till intromission occurs Pelvic oscillations followed by somnolent phase	Nine minutes Intrauterine	Many boars will serve a sow three to seven times in heat period
Sheep	30 seconds to five minutes	Noses vulva Genital olfactory reflex Paws with forefoot Bleating, stamping with forefoot, rapid licking Genital olfactory reflex	Very quick single pelvic thrust with forelimb clasping	Five seconds Intravaginal	Ram will sometimes serve estrous female several times. Some mature rams will serve each ewe only once

Clasping by the male during intromission and mounting is an important component of coital behaviour in most ungulates. The stallion and the bull effect tight clasping of the respective female with their fore legs adducted into her flanks. In the case of the bull, this clasping increases in intensity at the moment of penetration and ejaculation. Vigorous clasping also takes place in mating between sheep, but when the male and female are heavily covered by wool clasping is inevitably impaired to some degree. Recent studies suggest that more effective mating takes place in sheep when the female has been shorn before mating activities commence and if this is true the explanation is likely to lie in improved clasping by rams which in turn improves ejaculation.

Following ejaculation, male animals show a refractory period which is a state of sexual exhaustion. The state of sexual exhaustion is not principally a physical one, however, and refers mainly to the loss of stimulus value by the female. A quick return to mating behaviour is shown by male animals when they are given an opportunity to mate a new estrous subject.

It is the normal procedure among farm animals for repeated matings to occur with any given female. Stallions probably re-serve estrous mares five to ten times in each heat period and most rams are noted to remate with ewes three or four times. Bulls are seen to remate with estrous cows repeatedly perhaps on five or six occasions. Male goats have also been noted to remate one female on three to twelve occasions. Boars normally serve sows several times over a period of 24 to 48 hours. Among poultry repeated matings are frequent.

Variations occur in the degree of receptivity in estrous female animals. Some studies have shown that many ewes permit about six matings during each heat period. When competition between ewes exists for a limited number of rams, it appears that older ewes are usually more successful than maiden ones in obtaining repeated matings.

A number of reports show that natural matings have the effect of shortening the duration of estrus in cattle. It is reported that the period of receptivity in cattle is shortened by as much as eight hours when natural circumstances are provided and repeated matings take place. These studies also show that when some female animals are 'teased' with vasectomized males, the duration of estrus is slightly shorter than otherwise. This substantiates the modern view that estrus is not under endogenous control alone and that its manifestation is subject, in part, to environmental factors including biostimulation.

25 *Parturient Behaviour*

Many free-living species seek out remote or concealed sites for giving birth and there are also strong indications that many domesticated animals deliberately try to avoid the hours of supervision when allowing the birth process to take place. However, because they can be kept under close supervision, the behaviour of farm animals during parturition is relatively well explored.

The process of birth passes through three very definite stages; this is normally the case with species in which single births occur. But in those animals in which multiple births occur (pigs for example) it can be said that there are only two phases, the second and third stages being interchanged regularly. The first stage refers to the dilatation of the cervix and the associated behaviour of the animal. The second stage is the expulsion of the fetus itself. The third stage is the passage of the afterbirth or fetal membranes. These phases extend into the broader, more general behavioural periods of prepartum, birth and postpartum.

PRE-PARTURIENT PERIOD

The pre-parturient period extends from late gestation (the carrying of the unborn fetus by the mother) to the beginning of the first stage of labour. Apart from certain changes in attitude towards any previous offspring still being nursed, there is generally little of significance in the animals' behaviour until parturition itself is very close. Once parturition is imminent, many animals separate themselves from the main group and select a site for the birth. Many species at free range choose inaccessible areas where the birth may occur unhindered. The domesticated ruminants often appear to withdraw from the grazing group when birth is only an hour or two away but, in some cases, the parturient animal has simply failed to keep up with the grazing drift of the main herd or flock.

In the immediate prepartum phase, the 24 hours before parturition, definite behaviour patterns begin to emerge. The animal becomes increasingly restless and frequently alters her position and possibly also her disposition. Gradually still greater restlessness becomes evident until a stage is reached where the animal changes her position every few minutes. Recognition of this pre-parturient behaviour in the pregnant animal allows the time of birth to be predicted accurately in the great majority of cases. Forewarning such as this allows the corrective husbandry of parturient animals and their neonates.

It has been observed in the pre-parturient sow that, while most of the time during the three days before the onset of labour is spent sleeping and feeding, an increasing amount of nest-building behaviour is shown. This is usually evident in the form of bedding chopping. A similar pattern of behaviour to that in the parturient sow is seen in the cow: much sleeping and feeding during the days immediately before parturition and moments of restlessness culminating in almost continuously restless and erratic behaviour. Sometimes, during the pre-parturient period, muscular contractions of the kind that herald the onset of labour itself may even take place so that one is given the impression that the actual birth is about to occur. There is strong behavioural evidence of the build-up of pain in the parturient animal during the late prepartum phase. The biological evidence of this pain in mares, cows and ewes has been considered extensively, and it seems that the pain serves to signal the forthcoming events to the animal. Pain secures the entire attention of the parturient animal and its total participation in the birth process. Increasing restlessness and other evidence of a build-up of pain constitute the predominant indications of the late prepartum period and the phase associated with the first stage of labour.

BIRTH

Pain is most evident during the phase which corresponds with the second stage of labour, i.e. birth.

As has been noted, the activities of the single-bearing (monotocous) dam are separable into three phases. In the multiple-bearing (polytocous) species, the parturient process is such that the activities of the second and third stages are interrelated, the two stages being interchanged at fairly regular intervals. It would be more accurate therefore to talk of only two stages, namely, the pre-parturient period and the period of fetus and fetal membrane expulsion.

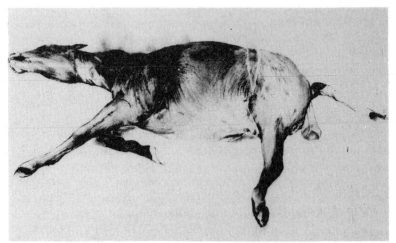

Fig. 82. A mare recumbent and in exertion during the second stage of foaling.

With the effacement of the cervix on completion of the first stage, the contents of the uterus can move through the open birth canal and invade the vagina. After the uterine contents have passed through the fully opened cervix the next phase of the birth can begin.

The outer fetal membrane remains adherent to the uterine wall and, during the course of physical straining, becomes rent with pressure. This allows the amniotic bladder containing the fetus to bulge the vagina—which is lined with secretions of cervical mucus and chorionic fluid—and effects further dilatation so that the fetus enters the pelvic canal. At this stage of labour the contractions of the uterus are regular. Even at the end of this stage, they can be very strong and frequent. These events terminate the second stage with an acceleration in the expulsive efforts of the dam. Provided there is no impediment to its delivery, the fetus is then expelled by a combination of voluntary and involuntary muscular contractions in the abdomen and uterus. Repeated straining, particularly abdominal straining, is therefore the principal feature of maternal behaviour at birth. The straining efforts increase in number and recur more regularly when the second stage of labour has begun. At this time the strong reflex abdominal and diaphragmatic contractions are synchronized with those of the uterus. The straining sessions are punctuated by resting intervals each lasting a few minutes. Further extrusion of the fetus is not necessarily achieved with each straining bout. The course of extrusion is subject to arrest and even retraction of the fetus back into the dam. One of the main obstacles to single births is the passage of the fetal forehead through the taut rim of the dam's vulvar opening. Once the head is born the rate of passage of the fetus is greatly accelerated. The shoulders follow the head within a few minutes and, immediately after this, the remainder of the neonate very quickly slips out of the birth passage. The mother's vigorous straining usually ceases when the fetal trunk has been born; often there is a short resting period at this point while the hind limbs of the neonate are still in the recumbent mother's pelvis (an event commonly occurring in un-assisted horse births). Immediately on being born, ungulate neonates exhibit typical struggling movements and upward tilting of the face before they make efforts to stand.

During birth, the posture of the dam varies a great deal. Some remain recumbent throughout birth; in others there is alternate lying, standing and crouching. The duration of the second stage of labour is usually much shorter than the first stage.

POST-PARTURIENT PERIOD

In the immediate postpartum period, the dam is engaged in the third stage of labour and in the grooming of the neonate (Fig. 83). The third stage is mainly concerned with the expulsion of the fetal membranes or afterbirth. There is not normally any vigorous straining during the expulsion of the membranes after their internal dehiscence. Fetal membranes are usually passed fairly effortlessly by the dam during the first few hours following the birth. Many animals occupy themselves by eating the afterbirth after its final expulsion (placentophagia). It has been said by some observers that the time thus spent by the mother parallels the efforts of the newborn in struggling and attempting to mobilize itself. Not all animals are placentophagic; cows and sows are, while mares are not. It seems that the two groups do have certain

Fig. 83. Mare commencing grooming of her foal immediately following birth (Przewalski horses).

general behavioural characteristics which differ. The species that are placentophagic usually keep their newborn close to the birth site for several days at least, whilst those that are not lead their sucklings away from the birth site very early on in the post-parturient period. It has been noted that in natural circumstances mares foal at night and in the open so that, by daybreak, the foals can trot and have been led away by their mothers. It does seem, however, that the instinct for placentophagy has been modified significantly in domesticated species.

SPECIES PARTURIENT BEHAVIOUR

Mares

The first indications that a mare is nearly foaling can be seen in the swelling of the udder and teats which, in most cases, becomes apparent about two days before the birth occurs. Also at this time, a wax-like fluid is emitted from the teats, although this may occur weeks before actual foaling. About four hours before parturition sweating is evident at the elbows and on the flanks. The first sign of labour occurs when the mare becomes increasingly restless. She may perform circling movements, look around at her flanks, get up and lie down spasmodically and generally show signs of anxiety. At the onset of parturition feeding ceases abruptly. The mare rises and lies down again more frequently than before, rolls on the ground and slaps her tail against her perineum. Subsequently she adopts a characteristic straddling position and crouching posture, frequently urinating at the same time. The mare may also show *flehmen*, especially after the allantoic fluid has escaped with the rupture of the allantochorion about the end of the first stage of labour when extremely vigorous straining—typical of the mare alone—occurs for the first time.

Just before straining starts, an unusually high raising of the head is sometimes observed. But when straining begins the mare soon goes down flat on her side and the expulsive efforts become intensified. From the first signs of sweating it may be deduced that the first stage of labour, which lasts for about four hours, has begun, but false starts are not uncommon. After some straining the waterbag (amniotic sac) becomes extruded; within it one fetal foot usually precedes the other. The bouts of straining become more and more vigorous until the muzzle of the fetus appears above the fetlocks. Although the straining bouts at this period are very vigorous the amniotic sac does not rupture. Most of the delivery time is normally taken up with the birth of the foal's head. Soon after this the remainder of the foal, except its hind feet, is expelled from the vagina. The reflex head movements of the almost wholly born foal finally burst the amniotic sac; the foal begins to breathe and its further reflex limb actions may extract the remainder of its hind legs from the dam. Final expulsion of the legs may be caused by the mare rising.

The duration of this second stage of labour is, on average, about 17 minutes although in normal circumstances it may last anything from 10 to 70 minutes. Following the completion of birth mares often lie, in apparent exhaustion, for 20 to 30 minutes. As has previously been stated, mares do not eat their afterbirths although they do groom their foals. The usual length of time for the third stage is about one hour; following the extrusion of the fetal membranes the foal engages in suckling within an hour and begins to trot after about four hours. Mares generally foal at night and in the open if possible.

Cows

Until the first stage of labour has begun physical changes rather than behavioural ones are apparent in the pre-parturient cow. Ingestive behaviour is, however, reduced at the time that labour is about to start. It is also at this time that the animal begins to show regular periods of restlessness; feeding sometimes recommences between these periods. Eventually the restlessness gives way to behaviour which is very similar to that encountered in conditions of colic. The cow appears apprehensive, looking all round her and turning her ears in various directions for brief periods. At this period the cow will also perambulate excessively if she is at all able. She will also occasionally examine patches of ground and sometimes even paw loose litter or bedding as though gathering it into one spot.

The first stage of labour becomes apparent when the cow goes through the motions of lying down and getting up again. She may also kick at her abdomen, repeatedly tread with her hind feet, look around to her flanks and shift her position frequently. About this time, also, the cow begins to pass small amounts of faeces and urine at intervals while arching her back and straining slightly. Cows tend to show these bouts of slight straining earlier on in parturition than the other farm species. With time, the spasms of evident pain become better defined and more frequent. Finally they begin to appear regularly about every 15 minutes and each spasm lasts about 20 seconds. The spasms are manifested by several straining actions in quick succession. After some bouts of straining, the allantochorion or first waterbag is rent and a straw-coloured urine-like fluid escapes. After this there is usually a short pause in the straining and muscular contractions. This pause terminates the first stage of labour which may vary in duration from three hours to two days, though a period of four hours is a more common (modal) average time.

About one hour later the more powerful straining of the second stage of labour becomes evident and the amnion appears at the vulva. This time, straining occurs about every three minutes and lasts for about half a minute but grows more powerful and more frequent when portions of the calf, such as its forefeet, become extruded at the vulva. At this stage the cow either adopts the normal resting position or lies on her side. Her upper legs may even swing clear of the ground if she strains while lying flat on her side. Straining is virtually continuous until the head and trunk of the calf extrudes. With the birth of the anterior part of the calf most cows quickly rise to a standing position and the remainder of the

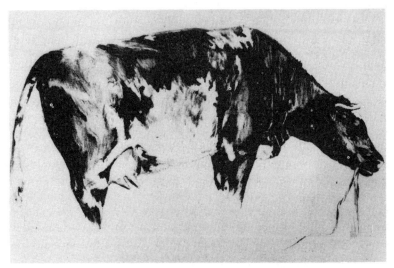

Fig. 84. Placentophagia in a cow following the third stage of calving.

calf (the hindquarters) slips from the pelvis; the birth is completed with the breakage of the umbilical cord. Occasionally a cow rising after the main period of extrusion will do so with the pelvis of the calf still lodged inside her own, and the retained fetus may swing from her for a period of time before dropping to the ground quite unharmed. Second stage labour, i.e. birth, is usually completed within an hour.

The cow often licks up her uterine discharges before the birth is completed. Once the calf is born, she rests for a variable length of time and then gets up and licks the fetal membranes and fluids from the calf. She usually eats the placenta (Fig. 84) and sometimes the bedding contaminated by fetal and placental fluids as well.

Ewes

It has commonly been observed that the ewe develops a premature maternal instinct. This is evident when the parturient ewe shows increasing interest in the lambs of other ewes. Occasionally ewes, evidently provoked by a particularly strong maternal drive, will attempt to take possession of other ewes' lambs and will sometimes do so successfully. It has been observed that well-fed ewes in good physical condition are more susceptible to this tendency than sheep in poorer physical condition, which would more likely characterize pregnant sheep under natural, undomesticated conditions.

Recent studies of the behaviour of parturient ewes have been made with a particular view to investigating the temporal aspects. The onset of parturition was first signalled in about 50% of ewes through physical signs: the protrusion of the amniotic sac and part of the fetus, the release of amniotic fluid and the passing of bloody mucus. In another 33% of cases, behavioural signs were at first more evident than physical ones. All the characteristic behavioural signs, even when they did not constitute evidence of impending parturition, were noted. Although no clear pattern was found, a picture has emerged in which the behavioural pattern of parturient pain is clearly evident. A small number of ewes separate themselves from the main flock when parturition is imminent. Most ewes display signs of nervous and restless behaviour: lying down and getting up again, paddling with the hind feet and other classic signs of discomfort. In 17% of ewes no initial signs of parturition were evident at all, although the sheep were kept under close supervision and almost constant observation at the appropriate times during two lambing seasons. Scraping the ground with a foot is common.

Although there may be frequent lying down and rising before birth, most ewes remain recumbent until the fetus is partially or completely expelled. In cases of twin and multiple births the neonates usually follow each other within a matter of minutes.

The mean joint duration of the first and second stages of labour is 80 minutes. The process of parturition in ewes would therefore normally seem to be a fairly swift one. The standard deviation in time of lamb delivery is about 50 minutes and the lambing times would follow the normal distribution were it not for cases of parturition of over 22 hours due to difficulty in birth (dystocia). There is no apparent difference in duration between breeds or ages. Although ewes lamb at all times during the 24-hour period, it has been observed over a period of time covering several lambing seasons that an unusual and disproportionately large number of ewes lamb during the four-hour period lasting from 7 p.m. until 11 p.m. and also during the early morning hours from 5 a.m. until 9 a.m.

Ewes may chew and eat parts of the fetal membranes but they do not consume the entire afterbirth.

Goats

The development of parturition in does and ewes is fairly similar. Immediately after birth, the maternal orientation of the neonate allows the doe to administer further, more intensive, grooming of the young. Mutual recognition by the dam and the neonate is important in this animal (and also in ewes). A doe will often reject her kid if it is taken away immediately after birth and returned after a lapse of time. In some cases it has been known for does to reject the young kid after only one hour of absence following parturition.

Sows

Prominent features in the pre-parturient sow include intermittent grunting, champing of the jaws and rapid breathing. There is also a significant enlargement of the mammary glands. During the 24-hour period before parturition nest-building also takes place. This activity may start up to three days before parturition. The nature of these nest-building activities may, however, largely depend on the material that she is provided with. The sow will attempt to clean and dry her selected birth site and will chew long grass or straw to provide bedding, carrying it a considerable distance if necessary. The location of the prospective birth site may be changed more than once. Pawing activities are evident where the sow uses her forelegs to move the bedding about. In general sows adopt and maintain a lying posture at rest before birth. It has been observed that free-living sows like to find a wooded area to build a den with dry vegetation. The dens are lined with chewed-up undergrowth and leaves and the site of the birth place is dry and warm, as far as possible. In a concrete pen the sow will still satisfy her nesting tendencies using any material that is provided for her use. She will often resist human attempts to disturb or relocate her nest. The amount of time taken over the building of a nest varies from one sow to another but they nearly all make use of the straw, hay and any other dry material that is at hand.

During the process of delivery the sow is normally recumbent and lies on her side, although there are occasions when some sows adopt a position of ventral recumbency (lying on the sternum). Vigorous movements of the sow's tail herald each birth and the piglets are expelled without a great deal of evident difficulty. Although, as has been stated, polytocous species more accurately exhibit only two stages of parturition, relatively little afterbirth is passed until all the piglets are born.

When farrowing is only a few hours away the sow alternately utters soft grunting noises and shrill whining sounds. As parturition approaches she begins to grunt more intensely and emits loud squeals. Farrowing more often occurs during the night than at any other time. During farrowing, the recumbent sow may occasionally try to stretch out and kick with the upper hind leg or turn over on to her other side. These movements force out fluids and a fetus may be expelled at this time. Sometimes, during the birth of a piglet, the sow's body trembles and if she is of nervous temperament she may emit grunting and squealing noises.

Piglets are expelled at an average rate of one every 15 minutes. Nervous sows often stand up after the birth of each piglet; this may be associated with temporary reduction in pressure in the reproductive tract. The entire farrowing process normally lasts about three hours. The sows pays little attention to her young until the last one is born and, when finally she rises, she sometimes voids quantities of urine.

The fetal membranes are expelled in batches of two to four after all the piglets have been born, although some small portion is often passed during the process. Many sows will eat all or part of the expelled afterbirth unless it is removed immediately.

Following parturition she will often call her litter to suck by emitting repeated short grunts and may emit loud barking grunts if an intruder disturbs the nest. The sow rarely licks or grooms her young but sometimes appears to try to position the piglets near her udder or draw them towards her teats using her forelegs in scooping actions.

It is important to provide close observation over a particularly nervous sow for it is in such an animal that cannibalism is most likely to occur. With such a sow also, the young are sometimes crushed by sudden and erratic movements. The piglets can be removed immediately after birth if the sow shows signs of dystocia. After the process is over and her piglets are returned she will usually display normal responses towards them.

TIMING OF PARTURITION

It is an established fact that more births take place in farm animals during the hours of darkness than at any other time. Over 80% of mares foal at night, for example, and evidence suggests that they may be capable of delaying parturition if they deem it necessary. Although it is accepted that sheep lamb at all times during the 24-hour period, an unusual number of ewes give birth in the early hours of darkness. More cows also calve in the night than during the day. However, this delaying activity cannot continue indefinitely; once the physiological processes of labour have progressed to a certain stage, the animal is no longer capable of postponing parturition.

Parturient synchronization points to a form of maternal protection which may be either conscious or unconscious. For example, it has been found that many mares foal during the early summer when nutrition is at its height.

Mares mated early on in the breeding season will often foal at the same time as those which mate later on in the season. Apart from this synchronizing effect on gestation, it has been found that seasonal factors may also affect the time of birth. Comparing two groups of mares, it was found that the average length of time of gestation was eight days longer in those which foaled in the spring than those which foaled in the autumn.

The birth times of lambs are synchronized to a degree; this is most evident in hill breeds. It has also been found in a herd of Angora does that, during one season, two-thirds of the kids were born within a four-day period.

26 *Nursing and Maternal Behaviour*

MATERNAL BEHAVIOUR

At the time of parturition the female subject acquires a novel repertoire of behaviours oriented towards acceptance and maintenance of the neonate animal. As in some other 'new role' situations the parturient animal has been behaviourally primed by an increased and special hormonal output. It appears that both the concentration of the reproductive hormones and their relative proportions to each other create the state of maternal behaviour.

At parturition most animals which normally are units of a flock or herd seek some degree of separation from their group when birth is imminent. This 'crypto-parous' characteristic allows the animal privacy to engage in the parturient processes delivering the young. Equally important, in this arrangement, is the state of seclusion which allows the immediate association between a mother and her newborn animal to develop during their own 'critical periods' which persist for several hours following parturition. The nature of critical periods has already been discussed in the section on learned behaviour. The pair, in comparative isolation, bond very quickly and efficiently together. It is of interest to note that when animals, such as sheep, are giving birth simultaneously under crowded conditions of confinement incidences of mismothering are common. In such circumstances some parturient ewes, already primed and into their critical period some hours prenatally, sometimes make strong attempts to steal alien lambs. In the maternal confusions resulting, many lambs may die.

Fig. 85. Mare grooming and cleaning amniotic fluid from newborn, wet foal (Thoroughbreds).

After the birth, the mother grooms the newborn in ways that are generally similar throughout the domesticated species of farm animals. The new mother removes the amniotic fluids covering the neonate by licking (Fig. 85). Neonatal heat loss through conduction, facilitated by thick amniotic fluid, is reduced. Much grooming activity becomes progressively directed to the dorsum of the neonate and to its head. These maternal attentions arouse the neonate and direct its primary interests towards its own mother. This system is not foolproof either in nature or in domesticated management. Some forms of the latter prevent this behaviour.

Following the grooming the mother has learned much of the identity of her young. Olfactory, gustatory, visual and auditory recognition has become established and will be reinforced thereafter. From this time on,

Fig. 86. Reciprocal stimulation between newborn calf and its mother as the neonatal bond becomes consolidated. (The animals are of a primitive breed of Steppes cattle in eastern Europe.)

the maternal subject will give care to the young animal and defend it with much intensity. The maternal attitude towards the young resembles recognition of the latter as, effectively, an extension of self. The young animal facilitates this by maintaining an intimate association with its mother, by vocalizing for assistance or support and by numerous nursing attempts.

The mother is motivated, in all her maternal behaviour, by a strong maternal drive. The dominance of this maternal drive can modify the normal reactivity of the subject, causing it to show changes in temperament. The maternal drive, if it becomes consolidated by a period of successful association with her young, will persist for a considerable duration; depending on the species, the maternal drive is likely to persist for months. Towards its termination reduction in drive intensity becomes apparent in the maternal behaviour. The mother shows increasing indifference to care-seeking (et-epimiletic) behaviour from her young.

Fig. 87. Close association between mares and foals after bond formation. On the left, the foal is maintaining physical contact (thigmotaxis) while it walks with its mother. On the right, the stationary mare oversees her foal turning its attention elsewhere.

In domestication some maternal drives appear to become pathologically altered by the duress of some forms of husbandry. Extreme aggression can be shown by newly farrowed sows towards attendants. This warrants cautious approach to such animals at this time, particularly if their piglets are to be handled. In some instances pathological maternal behaviour in sows takes the form of cannibalism upon their own litters. Immediate sedation of such cases is required. Many will show normal maternal behaviour on recovery from sedation.

Abnormal maternal behaviour is shown by a small proportion of individuals in all the domestic species. Much of this leads to rejection of the newborn animal while it is still in its critical period when imprinting should occur. Such events also deprive the newborn animal of intake of colostrum which also, for optimum passive immunity, should be ingested promptly after birth under stress-free conditions. It is believed that the aetiology of several neonatal diseases is complicated by such behavioural malfunction.

NEONATAL BEHAVIOUR

At birth the newborn animal shows immediate reactions to its novel environment. Among the farm animals the newborn characteristically acquires a position of sternal recumbency with the forefeet, extended during expulsion at birth, flexed to support the forequarters while the animal lies at rest for a short time. The neonate raises its head, moves it about, shakes it vigorously. The nostrils are cleared by head-shaking and snorting. At first the ears are limp, but within about 15 minutes, in neonates which are developing their postnatal reactivity normally, the ears have become erect and mobile. At this point the neonate is showing increasing interest in its immediate environment which, under natural conditions, is dominated by the presence of its mother.

With the attainment of adequate 'awareness' the neonate then struggles to attain an upright posture. It does so using patterns of movement which are typical of the rising motions of the adults of that species. Several attempts are normally made with increasing success until the animal has acquired the ability to stand upright. Normally this development has been achieved by about one hour postpartum. When the animal is upright its behaviour then becomes oriented towards major features of its environment such as a wall, an attendant, its mother. Under the best and most favourable circumstances the neonate's attention becomes increasingly focused on its dam. The animal begins to explore the ventral contour of the maternal abdomen and udder-seeking is then pursued. The newborn's direction, with the head and neck extended, tends to follow the underline in its upward direction. The usual contour of the mother therefore leads the neonate towards her inguinal region. Strategic but minor shifts in the maternal stance are sometimes made to facilitate this. When the udder is located the neonate has its attention more specifically directed to this part and grasping attempts are made by the lips to secure prehension of a teat. With a teat located the sucking reflex is immediately induced and ingestion has commenced. Thereafter this procedure becomes greatly improved by learning and within a few hours of birth the newborn animal is a proficient nursling. In some circumstances, perhaps through misleading conformation in the mother or through incompetent neonatal

Fig. 88. A three-week-old litter of piglets pressing demands for suckling on their sow, who responds by adopting recumbency in which her hindquarters are dropped last: a potential danger to smaller piglets in that location of the teat order.

behaviour, the teat-seeking process can be very protracted. The drive for this goal may become fatigued. The newborn animal may fail to find its source of colostrum and its survival prospects are obviously poor without skilled intensive care.

In the first day of nursing activities the neonate may feed hourly throughout the day and night. Large quantities of milk are ingested during the first day. The animal sleeps during most of the intersucking periods. As the animal becomes some weeks of age nursing may become less frequent. At this time, although most mothers will have accommodated their young on demand till then, the mother–offspring association may undergo some equalization in which the mother may not permit suckling in every instance. In the nursing sow it has been found that about 20% of suckling episodes with the growing litter are not associated with any let-down of milk. This leads to the recognition that some sucking is not nutritional in function but provides for alternative comfort-need. Nursing attempts are evident in comfort-need situations, particularly with growing young in situations of sudden alarm.

Part VI
Abnormalities in Behaviour

27 *Stress in Animal Husbandry*

Since normal behaviour can be shown by ethology to relate to relevant and complex circumstances, abnormal behaviour must also be capable of being shown to relate to their own specific circumstances. The preliminary stages in the development of diagnosis in animal disease investigations often are based on this truism. Some veterinary clinicians have emphasized the need to recognize, in a systematic fashion, this relationship between abnormal behaviour and the principal physical factors which operate causatively. Ambient factors can clearly operate in a causative role in abnormal behaviour.

Many ambient factors can be seen to affect behaviour by producing adaptive responses. Some ambient factors can affect behavioural properties making them manifestly abnormal in character. In these latter cases successful adaptive behaviour has not apparently occurred in response to the ambient factor. Ambient factors, in management for example, can be instrumental in the production of stress. These factors are stressors. Ambient factors causing abnormal behaviour must be singularly identified if they are to be appreciated and controlled. Pathogenic ambient factors may become as important in animal disorders in future as micro-organisms have been until now. Stressed animals are, by nature, predisposed to disease. Management practices evidently create many stressful circumstances, often by preventing or restricting the behavioural patterns inherent in animals.

'STRESS' AS A VETERINARY CONCEPT

Liberal use of the term 'stress' in a veterinary context led to confusion. It should be emphasized that the term 'stress', as used in a veterinary context, refers not to any single parameter or set of bodily reactions, but to a heterogeneous assortment of phenomena.

Stress has a considerable variety of meanings in physiology, psychology and human medicine. It has been used by veterinarians and agriculturalists in discussions of a host of husbandry problems, many of which bear little obvious relation to each other or to earlier conceptions of stress.

Historically, the question was raised by Darwin in his *The Expression of the Emotions in Man and Animals*. He pointed out that a variety of distinct experiences give rise to many of the same outward signs. The body may tremble, for example, during fear or anger, during exposure to cold, after severe injury or excessive fatigue and in various diseases. Clearly, Darwin realized that there are some important similarities among the reactions given by the body to a variety of external factors which today would be called 'stressors'.

Such observations found a physiological basis in the work of Cannon. Like Darwin, Cannon was impressed by the similarities of the bodily responses to a variety of disturbances; but Cannon looked deeper within the body, putting special emphasis on the activity of the sympathetic nervous system and the secretion of adrenaline. He noted that both these parameters are increased during emotional states and pain and that such increases account for a large number of physiological changes, including a rise in blood sugar, the improved contraction of fatigued muscle, the hastened coagulation of blood, an increase in the number of red blood cells and changes in the distribution of blood volume throughout the body.

Although Cannon used the word stress in discussing such findings, it was Selye who initiated the research which is commonly associated with the term. In the course of an endocrinological study Selye noted that crude ovarian extract, when injected into rats, produced a number of effects which are not normally associated with ovarian function. These effects included gastric ulceration and enlargement of the adrenal cortex. In following up this unexpected finding, Selye discovered that injections of a great variety of foreign substances produce the same set of changes; since these changes were evoked by so many different agents, Selye referred to them collectively as the general adaptation syndrome (GAS). The list of stressors which trigger the GAS was soon expanded to include wounding, exposure to cold and physical restraint.

Attention has now been given to physiological parameters which do not fall within the classical accounts of stress reactions. Studies of animals under duress frequently document changes in the liver, spleen and other organs, in muscle metabolism and in various secretions of the pituitary and thyroid glands. When the scope is thus broadened, more emphasis naturally falls in the fact that different stressors evoke many different reactions as well as similar ones. In fact differences emerge even when study is limited to aspects of the classical GAS: for example, extreme cold increases the secretion of the glucocorticoids by the adrenal cortex while extreme heat does not.

Another development, and one of fundamental importance for veterinary purposes, is the growing realization that animals mediate with their environments not only in order to feed, drink, copulate and avoid pain, but also to render a barren environment less monotonous, to gain access to social companions when housed in solitary confinement and to avoid extremes of temperature and illumination. We observe that while too much stimulation can act as a stressor, animals will also expend considerable effort in order to avoid too little. It is likely that some types of under-stimulation will have physiological effects which are quite the opposite of those caused by over-stimulation. Usually in a veterinary context, the term stress is used when there is a profound physiological change in an animal's condition, generally leading to a disease state. But even within this comparatively specific framework, the physiological pathways of stress prove to be highly diverse. A few examples may illustrate this diversity.

Perhaps most drastic is the 'acute stress syndrome' which accounts for a substantial number of cases of sudden death among pigs. The problem is most commonly associated with the transporting of animals but deaths can also apparently result from milder forms of handling and after such routine treatments as weighing and tattooing. In these cases, post-mortem examination reveal no infectious disease or physical injury.

Again, the mortality rate of pigs during transportation is related to the temperature and to other environmental factors or stressors to which certain animals are genetically more susceptible than others perhaps because of differences in endocrine functions.

A common condition which is said to reflect stress is the development of gastric ulcers. Gastric ulcers of pigs are not a consequence of any specific infectious agent, but can be precipitated by adverse social and environmental factors. Perhaps gastric ulceration in pigs represents a type of stress reaction similar to those studied by Selye.

Diarrhoea with coliform enteritis is often thought to be associated with stress. Neonatal calf scours can be precipitated by fatigue due originally to social and environmental disturbances during the first days after birth. Among pigs the condition is particularly common shortly after weaning; the highest incidence of the disease falls in the 20-day period after weaning, regardless of the age at which weaning is carried out. In these examples with both calves and pigs, the stressing factors appear to have lowered the animals' resistance to a specific pathogen.

These examples illustrate that veterinary conditions associated with stress are very diverse and almost certainly involve a number of quite different physiological pathways. In formulating a veterinary definition of stress it is important to avoid any implication that the term refers to some single phenomenon which, if sufficiently understood, would provide the clue to the treatment and prevention of a host of disease states. Instead a more general concept is required which can be used to refer to and emphasize the fact that many environmental and management factors can contribute to disease in a variety of subtle ways. In defining the term several important factors have to be considered.

There has been some confusion as to whether stress should refer to a bodily state or to the environmental agents which produce this state. Selye appreciated the need for clarity on this point. He used stress to refer to the reaction of the animal and employed the word stressors for the environmental agents involved. There is no apparent need to modify this usage for veterinary purposes.

Some individuals speak of stress as an undesirable extreme in an animal's condition, while others use the term to mean a normal and desirable parameter which sometimes rises too high or falls too low. The latter usage is so far removed from everyday senses of the word that it is likely to cause confusion. Equally important, this usage might be taken to imply that 'under-stress' and 'over-stress' can meaningfully be distinguished as opposites. It might be thought that in social crowding, animals are exposed to too much social stimulation, while in solitary confinement they are exposed to too little. Yet both these treatments can lead to increased activity of the adrenal cortex.

Another issue is related to Selye's distinction between the *stage of resistance* and the *stage of exhaustion* in the course of stress reactions. Should the term stress refer to the condition of animals which are managing to cope with adverse circumstances, or should it be reserved for conditions bordering on collapse? Certainly there are important biological distinctions between these two conditions, but both are central to veterinary problems and both can result from the same husbandry practices. It would seem useful, therefore, to follow Selye's lead in referring to stress in both cases.

There is disagreement whether the term stress should be used only when some clinical change has been demonstrated, or whether behavioural disruptions could by themselves be regarded as symptoms of stress. Certainly an animal can sometimes avoid serious physiological upset (e.g. hyperthermia) through altering the incidence of common behavioural patterns (e.g. wallowing, panting). But a system of husbandry is clearly in need of improvement if dire consequences are prevented only by gross changes in either behaviour or physiology. For practical purposes, then, states involving abnormal behaviour do deserve to be placed under the general heading of stress.

To summarize, we require that the term be defined sufficiently to be used in a tangible way in discussing a variety of veterinary problems. The term should be used where there are extremes of bodily states, but should not imply any measurable parameter which necessarily summates various reactions to adversity. Furthermore, the term should encompass states of coping as well as those of collapse and states involving disturbed behaviour as well as those involving altered physiological function.

With these considerations in mind, the following definition has been offered and generally accepted. An animal is said to be in a state of stress if it is required to make abnormal or extreme adjustments in its physiology or behaviour in order to cope with adverse aspects of its environment and management. A husbandry system can be said to be stressful if it makes abnormal or extreme demands on the animals. Finally, an individual factor, such as low temperature, may be called a stressor if it contributes to the stressful nature of a system of husbandry.

A feature of this definition concerns the issue of environments involving 'boredom' or physical restraint. Stress may arise from systems which make excessively high demands on the animals, as the definition states, but what about environments whose demands are excessively light? For example some species normally travel long distances for food in the wild. When such animals are kept in zoos, food is generally presented to them and this may be a major source of difficulty in the animals' adaptation to captivity.

Problems of this sort are actually covered by the definition. It is widely accepted that animals in monotonous and restricting environments seek out opportunities for exercise and stimulation. Most veterinary ethologists now suggest that the restriction of movement, 'boredom', thwarting of drives, stressful stimuli and deficiencies of the environment may lead to abnormal stereotyped behaviour. In this sense an environment may indeed cause an animal to make significant adjustments in its behaviour or physiology and would therefore be called stressful by the definition given above. Furthermore, the definition takes a stand on this issue of limited stimulation and freedom.

Table 8. *Environmental stressors associated with chronic control of livestock which may act cumulatively on animals*

Stressor origin	Stressor item
Management	Improvident welfare
	Nutritional level
	Husbandry standard
	Environmental variables
	Hygienic standards
	Noise level
	Attritive management policies
Space	Social density
	Peck order status
	Group size
	Permitted movement
	Area per head
	Isolation
Constraint	Hardware controls (stalls, tethers, races, crushes)
	Special suppressive devices
	Restrictive housing systems

An environment may seem, to a human being, to be extremely restricting and monotonous, but it cannot be called stressful unless the animals show abnormal behavioural or physiological changes in adapting to it. Furthermore, an environment may appear to be adequate and humane, while actually lacking some specific stimulus which is important for the animals' normal functioning.

Finally, it should be noted that this type of definition does not relate stress simply to a fall in production from an existing commercial level. It is a fair hypothesis that many important stressors will reduce the productivity of livestock below the maximum potential; but on many farms individual animals are already functioning below their maximum potential, often for reasons of economy, making deviations from existing efficiency levels unsensitive indicators of noxious environmental factors.

CONCLUSIONS

Whilst many environmental features are evidently stressors, it must be acknowledged that some specific environmental factors are putative stressors. This inconclusive status is a result of inadequate study, at either micro- or macro-analytical levels of study. For example, noise *per se* is not necessarily a stressor but increase in this stimulus in volume or duration may give it a noxious environmental property. The stress response to this is also uncertain at this time.

It is sometimes argued that a stress-induced behavioural anomaly (such as stereotyped movements) is a sign of adaptation. In other words, behavioural homeostasis has been effected to mitigate or eliminate the stress. This interpretation, in terms of animal welfare, is illogical. Those who argue that the perseverance of anomalous behaviour is adaptive may overlook the fact that the causal factors in the aetiology of such behaviour are still present. These, and other behavioural anomalies, may be attempts to achieve adaptation, but the adaptive plateau that is attained may be pathognomonic of stress if it persists and there is no restoration of normal behaviour patterns (i.e. dysstasis rather than homeostasis). Further decompensation may follow, both behavioural and physiological, affecting productivity, growth and disease resistance.

The chronological sequence in many of the anomalous conditions described show such a close and/or well-defined temporal relationship between stressor and anomaly that the connection between the two is clearly demonstrated. It is believed that very many others await recognition.

28 Anomalous Behaviour

Animal scientists are beginning to realize that many forms of abnormal behaviour are related to noxious stimuli or stressors in an animal's environment. Stress is a dynamic state with significant behavioural manifestations among which displacement activities feature very prominently. There is every prospect that more detailed study of anomalous behaviour in farm animals in the near future could improve the diagnosis of stress which, like so many other diseases, undoubtedly occurs with differing degrees of intensity and variations in the signs displayed. Already it has been learned that when the 'quantity' and 'quality' of an animal's environment are reduced there is an increased probability of abnormal behaviour developing. The decline in environmental quality includes a reduction in the variability of the animal's surroundings. Inferior environments of this type are closely linked with anomalous behaviour such as cannibalism, reduction in appetite, stereotyped movements, poor parental care, over-aggressiveness, unresponsiveness, tail-biting, cribbing, etc. Many of those behavioural features which have for many years been regarded as vices are in fact forms of anomalous behaviour resulting from exposure to environmental inadequacies.

The incidence of anomalous behaviour in animal husbandry appears to be on the increase. With modern animal production methods it is sometimes considered adequate to provide the animal with facilities for nutrition, rest and reproduction, but little else. The environmental facilities afforded to pigs, for example, are commonly very limited; yet field studies on pig behaviour have built up an ethogram for this animal along the lines already described in Chapter 22. Thus the free-ranging pig is commonly found to spend 40% of its time at rest, 35% of its time exploring and investigating features of its environment, 15% of its time feeding and drinking and 10% on sundry other activities. The modern intensive husbandry methods do not normally allow pigs to indulge in voluntary activities to this extent, and the inadequacies of environmental conditions are eventually revealed in overt anomalous behaviour by the animals.

However tedious it may be it is essential to lay down some definition that will divide normal behaviour from abnormal or anomalous behaviour so that the latter can be pinpointed more certainly. In general, if observations establish constancy in a specific feature of behaviour in the majority of animals under similar conditions, it can be taken that the behaviour observed is normal for that situation. However, more precise guides to classification of behaviour are required. Undoubtedly of prime importance is a decision as to whether observed behavioural features are appropriate in both their biological purpose and their degree of manifestation; inappropriate behaviour can be termed anomalous.

Modern forms of animal husbandry are progressively developing into systems which increase the density of animals held in groups. Comparatively little knowledge of the responses of animals kept under this type of management was formerly available. Farm animal behaviour studies have moved quickly to keep abreast of these newer methods of husbandry which involve animals being accommodated under conditions of limited space.

It may well be that, in addition to crowding, one of the stressors which contributes to anomalous behaviour of domesticated animals is a lack of diversionary quality in their environment. It is possible that animal producers have been insufficiently aware of the needs of farm livestock for 'quality' within the environment. So much attention has been focussed upon the hygienic needs of animals in recent years that the behavioural needs have evidently been very largely overlooked.

The gathering of some diverse observations on behavioural aberrations which are paradoxically both irrelevant and significant is necessary. A synthesis of anomalous behaviour in animals can be attempted now.

Accumulated observations now clearly show that appraisal and recognition of etho-anomalies are a sound and logical means whereby stress in animals can be identified. Biochemical or physiological data alone may be misleading or inconclusive. Behavioural homeostasis, for example, may adaptively help regulate and maintain normal physiological functions so that analysis of the latter may not reveal that the animal is under any apparent stress. Stressors can be much more easily identified since they are usually tangible features of the immediate environment of the animal. Some stressors have greater impact or significance than others.

In studying anomalous behaviour within husbandry systems, animals are seen to be capable of tolerating a sum of stressors up to a given level. The critical level varies with certain individuals within the population of a given type of animal. A critical level of stressors can be reached in a group which is capable of exceeding the effective adaptive responses of representative individuals in the group. The threshold level of stress tolerance of the composite group is then exceeded. At this point it is generally observed that stressors generate anomalous behaviour.

The feature of population density which gives it its noxious potential can be most clearly described by the simple term 'crowding'. Even in species of animals which have become very adaptable to conditions of high social density in conventional domesticity, a limit of ability to adjust to degrees of crowding which prevent a social hierarchical system from operating satisfactorily can readily be reached. The form of aggression exhibited between animals to maintain a stable peck-order is usually no more than a gesture. When the subordinate animal does not have adequate space to avoid the aggressor's gesture, the latter then takes the form of true agonistic behaviour. Resultant injuries and wide disparities in social ranking then become imposed on subordinates. Subordinates so marked then become subjected to increased aggressive attention. The total incidence of aggression within the group then escalates. In a comparatively short time injuries from agonistic encounters can be found throughout most animals in an affected group.

The two conditions of lack of quantity and inadequate quality can combine to increase the stressful character of some forms of animal control to the point where abnormal behaviour becomes an inevitable outcome. The range of these is much greater than those described in the previous section. Some specific conditions can be given to indicate the wider range of abnormal behaviours.

SPECIES FORMS OF ANOMALOUS BEHAVIOUR

Swine

When confined within stalls such as feeding stalls or farrowing stalls, for extended periods of time, swine frequently exhibit anomalous

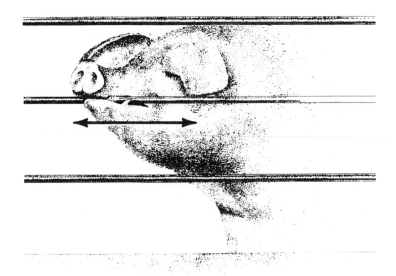

Fig. 89. Oscillative bar-mouthing in a chronically installed sow is of diagnostic value in determining stress.

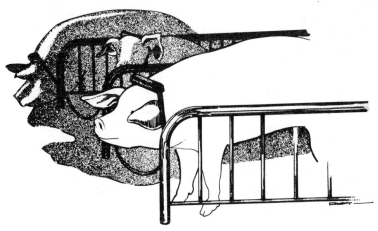

Fig. 90. Acute restraint in breeding sows in a modern system of husbandry involving stall enclosure, chained tethering and bedding-free slatted floor. While seeming to be labour-saving and economical, this type of husbandry is stressful and reduces long-term breeding potential, induces cases of 'thin sow syndrome' and generates anomalous behaviour in the form of stereotyped activities.

behaviour in the form of habitual mouthing of metal fixtures. Chronically, confined sows, for example, will engage for long periods of time on chewing upon steel nipple-type automatic waterers. They will also engage in bar-biting behaviour for extended periods of time, on galvanized piping at the front of their stalls. Crowded swine frequently show tail-biting. Cannibalism is another apparent consequence of environmental stress.

Cattle

In cattle, numerous forms of anomalous behaviour are evident in systems of management which feature close confinement. Among groups of young calves, for example, kept under crowded conditions within pens, intersucking behaviour can be observed. In this form of etho-anomaly, the calves are seen to indulge in excessive sucking of the underparts of their companion animals. The anatomical parts which are subjected to this excessive sucking include areas such as the navel, prepuce, scrotum, etc. In growing calves kept in confinement excessive self-grooming is frequently encountered. They may lick a pen fixture excessively, forming pools of saliva on the floor beneath these fixtures. If young weaned calves are maintained in dense groups, mutual grooming, to excess, can be observed. This excessive grooming can lead to the formation of hair-balls in the alimentary canal with such clinical consequences as acute obstruction.

Fig. 91. Tongue-rolling in cattle.

A recently noted form of anomalous behaviour has been observed in adult dairy cattle which are closely confined in pens during their non-lactating periods. This anomaly is in the form of 'tongue-rolling' (Fig. 91). The tongue is rolled up within the mouth, which is usually held open whilst this is done. The practice can be performed repeatedly. In some of these instances partial swallowing of the tongue is performed. Where this anomaly has been observed, it has also been noted that the animals are typically subjected to severe restriction of movement.

Sheep

In modern sheep husbandry there is a progressive tendency to maintain sheep in dense groups indoors. Among breeding ewes in forms of experimental husbandry which are being developed and which involve them being chronically controlled within pens of limited size, set out in rows within sheep houses, anomalous behaviour in the form of 'wool-picking' or 'wool-pulling' has been observed. This etho-anomaly takes the form of sheep pulling with their mouths on the strands of wool on the backs of other sheep around them. Ultimately all the sheep in the affected group lose long wool over patches of their back or even over the entire area of the back. While this wool-pulling is not carried out to the extent that bare areas of skin develop, only wool fibres of approximately 2 to 3 cm remain over the back while fleece of normal length is still carried by these animals on their sides and hindquarters. Ewes seem to select individuals and pick on them day after day.

Many breeders of young rams now mature these animals in large numbers in small paddocks. The animals are therefore subjected to conditions of very high population density. Among such rams it has been observed that there is a considerable amount of homosexual activity and the progressive development of homosexual drives. This becomes particularly noticeable at the age of puberty. After puberty, it is reported that many of these animals maintain this anomalous behaviour even when put into conventional breeding systems with female sheep. Many of these rams eventually adjust to normal breeding activity; some apparently do not.

Poultry

The innumerable reports of anomalous behaviour among poultry associated with crowding have become common knowledge. The principal one of these is cannibalism: a minor form of this is feather-picking. Cannibalism can be seen in adult poultry and it can also be seen in young poultry at the brooding and rearing stages. This condition has now become so commonly recognized that it appears to have justified the widespread routine debeaking of young birds, to limit but by no means prevent the consequences of this anomalous behaviour. Broiler poultry on deep litter may develop serious problems from ingesting large quantities of litter. Boredom and social facilitation may be involved in this pica.

Various forms of displacement activities have also been reported. These displacement activities, when they increase in intensity and incidence, can become anomalies. Excessive crowing has been noted among adult roosters maintained in long-term confinement.

Horses

The stable vices of horses are probably the forms of anomalous behaviour which have been longest recognized among the domesticated animals. These are well understood to be the consequence of boredom in horses due to their being kept within stables for long periods of time without the provision of adequate exercise. The most common forms of behavioural vices in the horse are probably cribbing and weaving. A horse is described as a cribber when it habitually sets its upper incisor teeth on a firm object such as a manger, and sucks in and swallows air, usually making a characteristic grunting sound at the same time (Fig. 92). In time, this addiction has a chronic adverse effect on the animal's health. Horses which have learned the vice of cribbing sometimes progress to another associated vice, wind-sucking or aerophagia. Wind-sucking is simply cribbing without the need of the horse to bite onto an object when the air is being sucked in.

Weaving in horses is a form of stereotypy or 'tic'. Stereotypes as with displacement activities can become etho-anomalies when their intensity and incidence reach excessive proportions. Weaving is a form of behaviour in which the horse will stand in his stall, rocking from one side to the other, or back and forward, in a repetitive, precise type of movement which usually involves stepping actions of the forefeet. Akin to this is stall-walking. In this latter form the animal may make use of the larger area afforded by a box, as distinct from a stall, and will engage in slightly more elaborated ambulatory actions than weaving. The identifying features are essentially similar being a highly repetitive form of stereotyped motion. Both of these etho-anomalies are often sustained for such durations of time that the drain on the animal's energy becomes significant. In these cases a deterioration in physical condition results together with painful back conditions from excessive lateral spinal flexion.

Fig. 92. Crib-biting with aerophagia in a chronically stabled horse. Note the vigour in this behaviour shown in the prominent neck muscles.

Another stable vice which has long been recognized in association with inadequate exercising and protracted maintenance within stalls is stall-kicking. Stall-kicking horses will kick quite forcefully with their hind feet against the sides of their stalls towards the rear or at rear posts of the walls. This kicking can be practised so frequently that the animals acquire chronic injuries to the lower parts of their hind legs.

Another etho-anomaly observed in the horse which is subjected to unvaried conditions of control is known by the common term 'sourness'. The sour horse displays an adverse deterioration in temperament, passes faeces which contain a significant proportion of material which has not been digested, participates reluctantly in activities for which it has been adequately trained and loses its general functional soundness.

Polydipsia nervosa is also seen in some horses isolated and confined in stalls with water supplied *ad libitum*. Some will consume about 140 litres daily (45 litres being taken as a normal upper limit). Such large quantities can precipitate gastric or intestinal volvulus. An associated polyuria may be the first indication of the condition. The habit can be broken by controlled water supply and increased exercise.

Head-shaking is yet another tic shown by horses which have been subjected to unsatisfactory management. This stereotyped behaviour occurs in several different forms. The most common is repeat 'bobbing' of the head up and down. When the habit is established it is very difficult to remove. Lack of appropriate exercise is thought to cause the condition. Stereotyped behaviours, such as this, are considered to have a self-hypnotic value by which stress can be alleviated.

The eating of faeces may be normal behaviour in very young animals such as foals. When it is practised by adult animals, it is abnormal. This form of anomalous behaviour can be observed most often in adult horses. In these circumstances it is usually found in individuals which are under chronic control and are not provided with adequate diversionary activities.

ANOMALOUS SEXUAL BEHAVIOUR

Males

Sex drive in male farm animals is commonly referred to as libido. Excessive libido is seldom considered to be a behavioural abnormality in animals, but an abnormal reduction in the level of sex drive in male stud animals is by no means uncommon. Indeed it has been suggested that *reduced sex drive* constitutes one of the principal causes of infertility in male animals. Some forms of impotence in farm animals are undoubtedly related to physical defects such as arthritis or other orthopaedic conditions which affect the mobility of the animal; but in a great many instances impotence is not evidently related to a physical impairment. It is suggested that in many of these cases the specific defective behaviour is genetically determined since this form of impotence is not infrequently encountered in certain breeds.

Male sex drive certainly has a heritable nature; this has been shown in bulls. The low sex drive of impotent bulls results in these animals having abnormally long reaction times when required to mate with cows under controlled conditions. While 50% of bulls make a positive attempt at mating within two minutes of being presented to an estrous cow, the broad population of bulls has an average reaction time of approximately 12 minutes. Bulls with reaction times of 30 minutes or as much as one hour are clearly, to some degree, impotent. During these protracted reaction times impotent animals characteristically show none of the sundry features of sexual interest normally shown by bulls. Attention is not directed towards the female, for much of the time, and the behavioural indications of sexual arousal, such as pumping actions of the tail-head and nosing and licking of the perineal regions of the cow, are not shown.

One particular type of impotence in bulls has been termed *psychic impotence*; this has been described by many workers in the field of animal reproduction. An affected bull usually shows a strong sex drive and mounts readily. When mounted however, although appearing to engage in normal movements, he fails to cover the cow satisfactorily. His hind feet are not brought close to the hind feet of the cow and close genital apposition does not occur between the two. After some futile pelvic thrusting by the bull, the mating attempt ceases. Such episodes may be repeated many times without effect. These bulls have been found to be capable of ejaculating satisfactorily into an artificial vagina and can be used as semen donors. In addition their sexual behaviour often seems to return to normal after a protracted period of sexual rest. Psychic impotence has been observed in bulls of various breeds and age groups. No physical cause has yet been ascertained for this type of impotence.

Anomalous sexual behaviour can quite commonly be observed in immature male animals. *Inexperience of mating* can be responsible for apparently abnormal behaviour which commonly takes the form of lateral mounting. This is shown by young bulls, in particular, when they repeatedly make mounting attempts over the sides of cows and may exhaust themselves before any effective mating can be performed. Some bulls, it seems, require time to learn the chain of behavioural actions necessary for effective mating. It may well be that in many of these animals puberty occurs much later than is commonly supposed. Lateral mounting certainly seems to be more of a problem in animals that are put to stud very soon after they have attained the chronological age by which puberty is expected to have occurred.

Abnormal sexual behaviour termed *somnolent* is not infrequently found to be a cause of impotence, particularly in aged bulls. In this condition, an animal shows protracted reaction times during which he lays his chin on the hindquarters of the cow and stands still with eyes closed, making no positive mounting attempts. Fatigue of the sex drive has obviously occurred in these animals.

Females

The transient behavioural state of estrus during which the female animal consorts with the male and participates in mating activities is also subject to anomalous features. These range from subnormal estrous behaviour to abnormally increased manifestations of estrus.

In a significant percentage of animals the physiological state of estrus appears to occur without the normally associated behavioural features being manifested. This condition is referred to as *subestrus* or *silent heat* and is something of a major problem in farm animal breeding. It occurs, for example, in mares and is also quite common in cows in heavy lactation. Studies of this condition in cattle have reported an incidence ranging from 18 to 30% among animals that were physically normal. In one study it was noted that there was a significantly higher incidence of subestrus in animals that were low down in the social hierarchy of the herd. It has been suggested that sexual functions are as likely to be suppressed in conditions of stress in the female animal as in the male.

Even when estrous behaviour is revealed the level of manifestation is by no means constant. The opinion has long been expressed that certain breeds of cattle, for example, have a hereditary predisposition for weak behavioural signs of heat or estrus. Various researchers have noted, however, that the apparent intensity of the behavioural signs of estrus in cattle is seldom consistent from one estrous period to another even in the same animal, and it is now generally recognized that the repeatability of behaviour at successive estruses is low. In the mare, however, the display of estrous behaviour appears to be fairly constant for a given animal. In this animal also, limited displays of estrus are quite common. *Continuous* or *excessively frequent estrus* is a well-known disorder among dairy cattle. Today it is recognized that this condition invariably relates to cystic ovarian degeneration and for this reason it has been described in the preceding chapter.

ANOMALOUS MATERNAL BEHAVIOUR

Epimeletic behaviour is the care-giving behaviour of the dam towards her offspring and the several aspects of epimelesis are subject to dysfunction. The main forms of anomalous maternal behaviour in farm animals involve repulsion of the newborn animal by the mother.

In pigs cannibalism is the most dramatic manifestation of this type of anomalous behaviour but other less spectacular forms of abnormal maternal behaviour occur. These include failure to groom the newborn animal, refusal to suckle (Fig. 93) and aggressive displays directed towards the neonate.

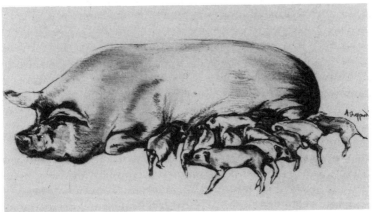

Fig. 93. Negative nursing response in a sow with a litter beginning to show the behaviour of hunger.

Various forms of anomalous maternal behaviour have been distinguished in sheep. Collectively these forms of anomalous behaviour contribute considerably to lamb mortalities. Examples include:

1. Premature onset of maternal behaviour. Aged ewes, in particular, are liable to show an abnormal interest in newborn lambs before the birth of their own. This can occur as early as two weeks prepartum. Such behaviour leads to confusion over lamb ownership and sometimes to later rejection of the ewe's own lamb when it is born.

2. Fostering. Some ewes, should they lose their own lambs, by death or other means, will make vigorous attempts to adopt alien lambs. Again, disputed ownership of a lamb does not enhance its prospects of survival.

3. Delayed grooming postpartum. Some ewes fail to groom their lambs at birth. A great many of those which give birth to twins or triplets fail to show satisfactory grooming towards the second or third born lamb. Ungroomed lambs remain wet and, in a cold environment, are more prone to suffer adversely from hypothermia.

4. Maternal desertion. Ewes, particularly young ones, sometimes desert their newborn lambs immediately following the birth. Deserted lambs usually die quickly if they have not been able to obtain any milk.

5. Butting of lambs. Aside from straightforward desertion, there may be a form of maternal–neonate disharmony in the form of aggressive behaviour directed by a ewe towards its lamb. The ewe, butting the newborn lamb away from her, very quickly discourages the teat-seeking chain of behavioural actions so necessary for the lamb's early learning of feeding behaviour.

6. Refusal to stand and facilitate teat suckling. Ewes of any age group may show this form of anomalous behaviour. The ewe typically edges away from the lamb when the latter is teat-seeking. For teat-seeking activities to take place successfully the mother has to remain stationary. In this condition teat-seeking or etepimeletic behaviour becomes exhausted in time.

Lamb-stealing in parturient ewes can represent a serious problem in sheep breeding. As a form of abnormal maternal behaviour, it is most typically seen when parturient ewes are crowded together. In this behaviour, ewes which have not yet lambed can be seen to make active attempts to acquire possession of newborn lambs from other ewes. As a consequence of this, lambs may become alienated and subsequently die of inanition.

THE SYNDROME OF PATHOLOGICAL ORAL ACTIVITY

This condition is a complex generic syndrome which contains a variety of manifestations of pathologically excessive and abnormally orientated mouthing behaviour in animals. Many of the manifestations have already been described above. They are presented here collectively to demonstrate that, while behavioural abnormalities may vary between species, age groups, and circumstances, they can be manifestations of a common basic behavioural disorder. Most forms of this syndrome are associated with the joint circumstances of chronic control on the one hand and hypostimulation on the other. Examples are tabled with other

Table 9. *Generic classification of 27 syndromes of anomalous behaviour in farm animals*

Anomalous cases	Species	Syndrome
Pathological oral activity	Horses	Crib-biting Polydipsia Aerophagia Coprophagia Wood-chewing
	Cattle	Excessive grooming Intersucking Tongue-rolling Pica
	Sheep	Wool-pulling
	Pigs	Tail-biting
	Poultry	Feather-picking
Stereotyped replication of motor activities	Horses	Weaving Stall-walking and pacing Head-nodding
	Pigs	Static and oscillative bar-mouthing
	Poultry	Displacement activities
Immobility and locomotor arrest	Mixed	'Freezing' (catatonia) 'Downer' (hypotonia) Balking (resistance)
Deviant perinatal behaviour	Sheep	Lamb stealing
	Mixed	Avoidance of neonate Agonistic orientation to neonate Refusal to suck Refusal to suckle Generalized puerperal aggression
	Pigs	Cannibalism

Table 10. *Some forms of pathological oral activity and their sequelae*

Examples	Clinical sequelae
Crib-biting in horses	Loss of condition, occasional colic
Tail-biting in swine	Loss of condition, abscessed hindquarters
Bar-biting in swine	Reduced production and subfertility
Intersucking in calves	Hairball
Excess grooming in calves	Hairball
Wool-pulling in sheep	Loss of fleece
Tongue-rolling in cattle	Unknown, possibly reduced physical condition
Feather-pecking in poultry	Loss of feather cover, trauma

generically classified anomalies which are adequately described by name.

RESTRICTION OF STIMULUS

In reviewing the above phenomena, it is clearly a scientific fact, that stress alters behaviour. The dose–response relationship, however, still requires critical study and evaluation. The environmental circumstances, which have been considered as forms of chronic control or chronic restriction, impose two main functional deficiencies of physiological perception. These are hypostimulation and hypokinesthesia.

Hypostimulation

Stimuli are perceptible external factors. Stimulation is the excitation process in the sensorium of the perceiving animal. The quantity of stimuli can affect the quality of stimulation. This is the basis of environmental quality in regard to its potential value in stimulation. Constant stimuli lose potential value in stimulation. A deficit in one type of stimulus can be compensated by an increase in another alternative stimulus, i.e. by variability. Stimuli contribute quantitatively to the quality of pooled stimulation through this phenomenon of compensatory variability. The alternating stimuli caused 'gating'. This colloquial term is used in neurophysiology to describe the phenomenon in which neutral 'action potential', which can be registered by electroencephalography as evidence of stimulation, shuts off in the presence of an extant stimulus and is restored when the animal's attention is switched to another and novel stimulus. The frequency of 'gating' increases with stimulus variability. Such variability is dependent on quality of stimuli and thereby affects quality of stimulation. Environmental quality is therefore a general description of an ambience and its possession of potential stimulus variability. Gating has an anxiolytic property. When this property is absent, anxious states can become established in the animal. These become manifested in etho-anomalies.

Hypokinesthesia

Abnormally decreased mobility and abnormally decreased motor function or activity is termed hypokinesia. This condition is known to have multiple causes. The principal effect of this is a marked reduction in the sensation of movement by the animal. It is this reduction in sensation of movement which is the critical factor.

The sense of muscular effort that accompanies a voluntary motion of the body is termed kinesthesia. The various forms of sense organs that respond to mechanical action, movement, position, touch and pressure constitute a major part of the sensory function of animals. The sense organs in tendons, joints and muscles collectively are a single group

Fig. 94. Hypotonia in a female goat subjected to restraint for breeding. The state of tonic immobility, once induced, will persist for a variable period of time in spite of aversive stimulation.

which are critically involved in the reception of kinetic stimuli and supplying a significant quantity of sensory input on a fairly continuous basis. The tactile senses respond to mechanical bodily stresses due to gravity acting on the body variably, according to its movement and position. These are only examples of the highly complex on-going input of the sensory system which supplies stimulation to the level of consciousness in the animal as a result of kinetic variability and suitability. Dysfunctions of this system, through reduced input or hypokinesthesia, are evidently capable of acting with adverse effect on functional mediation, so as to generate anomalous forms of behaviour.

It can be seen that many anomalies in behaviour can result in animals being impaired in terms of production. There is an ever-increasing realization that abnormalities in behaviour are early-warning signs that an animal's environmental circumstances are unsatisfactory. By pinpointing deficient environments in this way it may be possible to prevent stressors being applied to farm animals to an ever-increasing extent. With the development of applied ethology it is hoped that stress may be prevented from becoming the principal disease complex of animals under modern conditions of husbandry.

29 Behaviour and Clinical Disorders

There are numerous references to the behaviour of sick animals in the classical literature of Ancient Greece and today it can be said that all practising veterinarians rely heavily on behavioural observations in aiding arrival at a correct diagnosis of ill health. A scientific interest in the altered behaviour arising from changes in health, i.e. the objective study of behaviour in diseased animals, is termed clinical veterinary ethology. We are, at present, at a very early stage in the study of this subject. There is a good deal of knowledge of a practical nature relating to this subject and it is desirable now to organize the more academic aspects of this branch of ethology.

Since it is possible, in many circumstances, to relate behaviour to physiological mechanisms, it should also be possible to relate behaviour to pathological conditions. Furthermore, specific behaviour patterns may be associated with pathological processes which do not reside in the nervous system, although it must be recognized that many general diseases, by extension, ultimately involve the nervous system.

As the study of animal behaviour becomes more precise, more demanding and more sophisticated, it becomes clearer that the relationship between behaviour and physiological processes must be emphasized. Likewise, veterinary work dealing with animals that have some organic malfunction must attempt to recognize and relate abnormal behaviour patterns with pathological conditions of the animal body and to do so in a systematic fashion. This has not been done in the past since its necessity was not appreciated. If it is important to be systematic about the connection between normal behaviour and normal bodily function, it is no less important to relate abnormal behaviour with bodily malfunction. One of the simplest systematic approaches is the anatomical one, in which an attempt is made to relate observed behaviour to a pathological condition of anatomical body systems such as skeletal, muscular, locomotor and digestive systems. This is probably the method adopted by the majority of veterinary surgeons. One obvious difficulty in this approach, however, is that pathological processes probably all involve more than one body system; this causes major difficulties in correlating specific behaviour and signs of disease with symptoms of a specific disease.

In spite of the importance attached to abnormal behaviour in clinical veterinary knowledge, the descriptions of abnormal behaviour in veterinary literature are generally couched in subjective terms. Such descriptions give little indication of how abnormalities in behaviour differ quantitatively from normal behaviour. Veterinary literature dealing with clinical conditions is peppered with such expressions as 'stupid posture', 'painful expression', 'dullness', etc. These terms do nothing to convey meaning to uninitiated readers. Moreover they do very little to enhance the scientific nature of clinical veterinary literature. For the future, one of the foremost aims of the clinical veterinary ethologist must be the accurate measurement and description of the temporal and spatial organization of abnormal behaviour in his patients. By this means, clinical veterinary ethology could develop a more penetrating clinical acumen and lead to a deeper understanding of animals in a diseased state and more efficient information transfer in teaching situations.

One must hasten to acknowledge the fact that the present state of knowledge of the behaviour of animals is largely satisfactory at the practical level. The veterinarian working under practical and field conditions inevitably becomes aware of many aspects of farm animal behaviour. But this type of practitioner increasingly needs and uses behavioural knowledge and requires the expertise necessary for the accurate assessment of the situation. This knowledge in the main relates to the handling of animals, their breeding activities, the behavioural signs of illness and also to their habits, all of which can lead to hazardous situations in the spread of diseases.

BEHAVIOUR AS A DIAGNOSTIC AID

While the behaviour of a normal and healthy farm animal is clearly the concern of many people, it is primarily the veterinarian who is required to understand abnormal behavioural activity. Increasingly the veterinary profession has its attention directed towards conditions of distress, discomfort, probable pain and deprivation in the farm animals. These conditions are stressors and stress is a disease with many manifestations. Appreciation of this fact makes the veterinarian better able to guide animal users and producers in the optimum and acceptable conditions of maintenance for farm animals in modern husbandry. This newer role for the veterinarian depends largely on the development of veterinary ethology and the use of abnormal behaviour as an aid to diagnosis. At the present stage in the development of veterinary ethology it is impossible to deal with this topic comprehensively. All that can be done is to consider a number of clinical circumstances where diagnosis can be established on the basis of an animal's behaviour. Even this assumes a full and accurate knowledge of the normal behaviour of that type of farm animal on the part of the observer.

Examinations of groups of animals, to appraise their behaviour, is sometimes undertaken as part of a welfare assessment. In such circumstances, the observer searches for abundant evidence of normal social interactions between animals and any evidence of anomalous behaviour in individuals. Behavioural examinations are best performed in a quiet space or enclosure with limited light where distractions will be minimal. Tranquillization should be avoided, obviously. Parts of the examination which may tend to excite the animal should be postponed until the end. At the conclusion of the examination it may be apparent that further specialized clinical tests are required, such as radiology, clinical pathology, neurology and specific medical examinations.

In determining the suitability of an animal for a given role, its characteristic behaviour is of paramount importance. For this to be determined with professional competence, specific berhavioural responses and performances should be known.

In the course of convalescence, the behaviours of self-maintenance return to the animal's behavioural repertoire. An extended period of behavioural examination of the animal in its bedded premises is usually necessary for convalescence to be appraised by this method.

Postural Behaviour

The postural behaviour of animals is one of the commonest behavioural features to undergo change in diseased conditions. It is therefore essential to appreciate normal posture as a basis for recognizing postural abnormalities for clinical purposes. The following are the main circumstances under which animals adopt abnormal postures:

1. Mechanical conditions involving loss of support or stability by the animal.
2. Nervous conditions in which there is a reduction in adequate neural function to maintain muscular tone.
3. Painful conditions which make it impossible for the animal to maintain customary posture.
4. Permanent adaptive changes which the animal may have acquired as a result of prior experience of any of the circumstances mentioned above.

Mechanical conditions influencing postural behaviour are many and the following few examples are given as illustrations. Fracture of the metacarpus in the horse makes it impossible for the animal to take any weight at all on the affected leg. Fracture of the humerus also leads to lack of mechanical support and a grossly altered posture. Severance of the flexor tendons in the horse leads to a sinking of the fetlock and a turning up of the toe. Spastic paresis of the leg in cattle results in a contraction of the gastrocnemius muscle as a result of which the affected limb becomes shorter. Congenitally contracted tendons in foals also make normal posture impossible.

Nervous conditions which can create abnormal postures include radial paralysis in the horse following prolonged recumbency during anaesthesia, for example. A lesion in the cervical vertebrae causes the condition of wobbler in the horse the main characteristic of which is a stiff neck. Abscessation of the lumbar vertebrae can cause it to adopt the 'dog-sitting' position for lengthy periods.

Painful conditions which cause abnormal postural behaviour include suppurative arthritis and osteomyelitis of fetlocks. The latter condition causes a tucking-under of the hind legs. Gonitis (inflammation of the stifle joint) occurs in horses, principally, causing them to point the ground with the toe of the affected hind limb.

Permanent adaptive changes may arise in a condition such as laminitis which can occur in all the hooved animals; those which have experienced laminitis for some period of time sometimes learn to walk on 'tip-toe' with the forelegs. This position appears to minimize pain. The adoption of this posture also means that the hind legs of the animal are brought further forward beneath it. Spinal abscessation in the pig may be the result of tail-biting and this may cause the posture of a hind leg to be altered. The common condition of foot-rot in sheep can lead, in some cases, to a state of osteomyelitis. In this condition, the affected animal frequently adopts a kneeling posture. Cattle which are kept in stalls and have experienced a form of chronic laminitis sometimes learn to stand back in the stall so that their heels overhang the standing. This posture allows the animal's weight to be transferred to its toes, thereby reducing pain. Cattle which have suffered acute pain in both medial digits may stand with forelegs crossed to take all the weight on the lateral digits. This condition sometimes occurs in abscessation of the sole and fracture of the third phalanx of one leg. In cattle it is sometimes noted that the limbs are advanced and rotated outwards taking the weight off the lateral digits. This condition is referred to as 'wing shoulder'. The condition is found in various states and sometimes appears to be due to laminitis and at other times is related to aphosphorosis. Cattle with wing shoulder, when made to walk, do so apparently normally. Other behavioural postures include a tripod form of stance when one leg is shortened, for example, in spastic paresis. Cattle which have experienced pain in their feet for some period of time sometimes lie with their hind legs extended out behind them. This appears to relieve pain in the feet.

In a study of posture as an aid to diagnosis it must be remembered that many postural abnormalities are not shown unless the animal is at rest in its usual environment. For this reason, patient and quiet observation of the animal may be necessary before abnormalities of posture can be detected and appreciated.

Reflexes

Reflexes should not be studied as isolated phenomena but as actions of the whole animal. Part III of this book dealt with some normal animal reflexes; mutual grooming and stretching after rising are probably the two most common simple reflexes seen in healthy stock. Several factors, including illness in general, can inhibit grooming and stretching reflexes. In cattle another reflex, tonguing of the nostrils, may be inhibited during illness. It has also been suggested that the eructation reflex in ruminants becomes inhibited in many illnesses and, as a consequence of this, distension of the rumen develops and becomes painful. This further inhibits eructating reflexes and leads to the condition of bloat which is seen associated with various illnesses in ruminants. Recognition of these minor reflexes in the normal behaviour of cattle is an indication of sound health and, consequently, their absence suggests that health is impaired.

Pain

The farm animals manifest pain in various ways, depending on the site and the severity of the pain and also on the temperament of the individual animal. The observation of pain and the identification of its source obviously plays a very important part in clinical diagnosis.

Collectively the signs of pain in animals give an impression of uneasiness, but the behaviour of an animal in pain has certain specific features which are recognizable. The facial expression of an animal in pain is often quite characteristic; usually there is a fixed stare within the eye. The eye is not as mobile within its orbit as in the healthy animal. The eyelids tend to be slightly puckered. The ears of animals in pain, notably horses, are usually held slightly back and fixed in that position for long periods. Animals suffering pain usually have dilated nostrils. These facial signs collectively give an animal a 'worried' expression (to

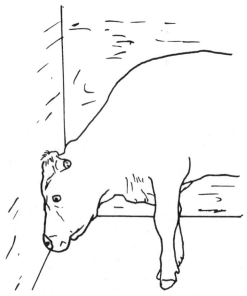

Fig. 95. A bullock showing 'head-pressing'.

use a subjective term). In pain the animal is often seen to turn its head to one side or the other, looking at one or other flank.

In colic or abdominal pain, the animal and the horse in particular shows various abnormalities of posture. Animals with persistent colic may show unusual recumbent behaviour; at other times they may adopt an unusual stance. Horses may back into a corner of a loose box and both horses and cattle can sometimes be observed standing pushing their heads against a wall (Fig. 95) when a painful condition is present in the abdomen. Abdominal pain may cause the animal to lie down frequently, rising repeatedly after short intervals. In between these periods of recumbency, a horse with colic may scrape at its bedding with a forefoot, whilst slowly pivoting around on its hind legs.

In conditions of severe pain, animals often show a full extension of the nostrils, rolling the eyes in the head, extending the head and neck vigorously and groaning. Some horses lie on their backs in a position of dorsal recumbency with all four legs held in the air. This abnormal posture may be maintained for up to 15 minutes. More violent manifestations of pain are shown by horses on some occasions: the animal may throw itself down, may roll from side to side, may rise and walk into objects in its surroundings. In this state the horse seems oblivious to its surroundings and all of its behaviour is indicative of severe pain.

Painful conditions of the skeleton frequently result in changes of posture in ways that have already been described above.

CONCLUSION

Over the past decade, much more has been discovered about the major divisions of ethology than could have been expected, even by optimists, to have been produced in such a period of time. Some areas of applied ethology have been particularly productive; such areas are social, sexual, perinatal, parturient, maternal, paediatric, feeding and sentient behaviour. Studies on social behavioural systems, in particular, have broken vast ground. Specialisms, such as ontogeny, ethogramology, stress, environmental responses, etc., have been rewarding areas of debate and study. The translation and absorption of the phenomena and the fundamentals of pure experimental animal behaviour studies into applied ethology is a continuous endeavour which is very reassuring when successful.

Many areas of livestock ethology have now been aggressively explored but there are still others where the first investigative efforts have still to be made. Perhaps it would be a fair comment to add that the latter are astonishingly numerous, so as to temper the enthusiasm of contemporary supporters of this profitable conceptual discipline of biology.

Modern forms of animal husbandry are progressively developing into systems which increase the density of animals held in groups. Animal husbandry is rapidly moving in this direction in spite of the fact that there is comparatively little knowledge for guidance on the responses of animals kept under this type of management. Veterinary ethology must therefore move quickly to keep abreast of the newer methods of husbandry which involve animals being accommodated under conditions of which the subject has no previous experience. However, veterinarians now regard objectivity—the pursuit of hard data—as a major goal in their work, but since it is considered that some element of subjectivity will necessarily be involved, symptomatology remains to be explored. For the latter to be done, clinicians must give increased attention to the subject of behaviour.

In spite of a rather late arousal of formal interest, rapid progress can be made in the accumulation of information on this subject. With effective means of harvesting some of its considerable practical experience of animal behaviour, and with encouragement for research and observation, the profession will make its vital contribution to this relatively new area of biology.

Supplementary Reading

Alexander, G. (1977) Role of auditory and visual cues in mutual recognition between ewes and lambs in Merino sheep. *Appl. Anim. Ethol.*, *3*, 65–81.

Alexander, G., Lynch, J.J. and Mottershead, B.E. (1979) Use of shelter and selection of lambing sites by shorn and unshorn ewes in paddocks with closely or widely spaced shelters. *Appl. Anim. Ethol.*, 5, 51–69.

Alexander, G. and Shillito, E.E. (1977) The importance of odour, appearance and voice in maternal recognition of the young in Merino sheep (*Ovis aries*). *Appl. Anim. Ethol.*, *3*, 127–135.

Angevine, J.B. (1975) The nervous tissue. In *A Textbook of Histology*, ed. W. Bloom and J.W. Fawcett, 10th ed., pp. 333–85. Philadelphia: W.B. Saunders.

Arnold, G.W., Boundy, C.A.P.., Morgan, P.D. and Bartle, G. (1975) The roles of sight and hearing in the lamb in the location and discrimination between ewes. *Appl. Anim. Ethol.*, *1*, 167–176.

Arnold, G.W. and Maller, R.A. (1977) Effects of nutritional experience in early and adult life on the performance and dietary habits of sheep. *Appl. Anim. Ethol.*, *3*, 5–26.

Arnold, G.W. and Pahl, P.J. (1974) Some aspects of social behaviour in domestic sheep. *Anim. Behav.*, *22*, 592–600.

Baldwin, B.A. (1975) The effects of intraruminal loading with cold water on thermoregulatory behaviour in sheep. *J. Physiol., Lond.*, *249*, 139–152.

Baldwin, B.A. and Ingram, D.L. (1967) Behavioural thermoregulation in pigs. *Physiol. Behav.*, 2, 15–21.

Banks, E.M. (1964) Some aspects of sexual behaviour in domestic sheep, *Ovis aries. Behaviour*, *23*, 249–279.

Beilharz, R.G. and Mylrea, P.J. (1963) Social position and behaviour of dairy heifers in yards. *Anim. Behav.*, *11*, 522–527.

Blockey, M.A, de B. (1979) Observations on group mating of bulls at pasture. *Appl. Anim. Ethol.*, 5, 15–34.

Bouissou, M.-F. (1978) Relations sociales chez les bovins domestiques dans les conditions de l'élevage moderne (Social relationships between domestic cattle under modern management techniques). *Proc. 1st Wld Congr. Ethology appl. Zootechnics*, *1*, 267–274.

Brantas, G.C. (1978) Ethological evaluation of cage- and sloped wire floor management for hens (Ethologische Beurteilung der Käfig- und Gitterroshaltung für Hühner). *Proc 1st Wld Congr. Ethology appl. Zootechnics*, *1*, 251–256.

Bryant, M.J. (1972) The social environment: Behaviour and stress in housed livestock. *Vet. Rec.*, *90*, 351–360.

Cabanac, M. (1974) Thermoregulatory behaviour. *Physiology*, Series I, vol. 7, *Environmental Physiology*, ed. D. Robertshaw. London: Butterworth.

Callear, J.F.F. and van Gestel, J.F.E. (1971) An analysis of the result of field experiments in pigs in the U.K. and Ireland with the sedative neuroleptic azaperone. *Vet. Rec.*, *89*, 453–458.

Christian, J.J. and Davis, D.E. (1964) Endocrines, behavior and populations. *Science, N.Y.*, *146*, 1550–1560.

Clutton-Brock, T.H., Greenwood, P.J. and Powell, R.P. (1976) Ranks and relationships in Highland ponies and Highland cows. *Z. Tierpsychol.*, *41*, 202–216.

Collery, L. (1978) Social interaction in an equine herd (Interacción social chez un groupe mixte de chevaux). *Proc. 1st Wld Congr. Ethology appl. Zootechnics*, *1*, 87–91.

Cowan, M.W. and Cuenod, M. (1975) *The Use of Axonal Transport for Studies of Neuronal Connectivity*. Amsterdam: Elsevier.

Craig, J.V. and Bhagwat, A.L. (1974) Agonistic and mating behavior of adult chickens modified by social and physical environments. *Appl. Anim. Ethol.*, *1*, 57–65.

Dudziński, K.L., Low, W.A. and Arnold, G.W. (1978) Ethological parameters as an index to ruminant–environment interaction. *Proc. 1st Ethol. Congr., Madrid*, 2, 22.

Dudziński, M.L., Pahl, P.J. and Arnold, G.W. (1969) Quantitative assessment of grazing behaviour patterns of sheep in arid areas. *J. Range Mgmt*, *22*, 230–235.

Ekesbo, I. (1978) Intensive husbandry methods as a contribution to stress and disease of farm livestock (Intensive Tierhaltung als Ursache des Stresses und der Krankheit bei Haustiere). *Proc. 1st Wld Congr. Ethology appl. Zootechnics*, *1*, 93–97.

Ewbank, R. (1964) Observations on the suckling habits of twin lambs. *Anim. Behav.*, *12*, 34–37.

Ewbank, R. and Bryant, M.J. (1972) Aggressive behaviour amongst groups of domesticated pigs kept at various stocking rates. *Anim. Behav.*, *20*, 21–28.

Fält, B. (1978) Differences in aggressiveness between brooded and non-brooded domestic chicks. *Appl. Anim. Ethol.*, *4*, 211–221.

Feist, J.D. and McCullough, D.R. (1976) Behavior patterns and communication in feral horses. *Z. Tierpsychol.*, *41*, 337–371.

Fraser, A.F. (1968) *Reproductive Behaviour in Ungulates*. New York: Academic Press.

Fraser, A.F. (1974) The dynamics of the unborn calf. *Livestock Int.*, *1*, 20–21.

Fraser, A.F. (1976) Some features of an ultrasonic study of bovine foetal kinesis. *Appl. Anim. Ethol.*, *2*, 379–383.

Fraser, A.F. (1977) Fetal kinesis and a condition of fetal inertia in equine and bovine subjects. *Appl. Anim. Ethol.*, *3*, 89–90.

Fraser, A.F. (1978a) The behaviour of maintenance (Editorial). *Appl. Anim. Ethol.*, *4*, 299–300.

Fraser, A.F. (1978b) A general review of sexual behaviour in livestock (Eine generelle Überprüfung des Sexualverhaltens bei Nutztieren). *Proc. 1st Wld Congr. Ethology appl. Zootechnics*, *1*, 507–512.

Fraser, A.F., Hastie, H., Callicott, R.B. and Brownlie, S. (1975) An exploratory ultrasonic study on quantitative foetal kinesis in the horse. *Appl. Anim. Ethol.*, *1*, 395–404.

Fraser, A.F. and Herchen, H. (1962) Tonische Immobilitat bei Tieren und ihre Bedeutung für die Veterinarmedizin. *Wien. tierärztl. Mschr.*, *49*, 271–276.

Fraser, A.F. and Herchen, H. (1978) The ultrastructure of behaviour (Editorial). *Appl. Anim. Ethol.*, *4*, 103–108.

Fraser, A.F. and Terhune, M. (1977a) Radiographic studies of postural behaviour in the sheep fetus. I. Simple fetal movements. *Appl. Anim. Ethol.*, *3*, 221–234.

Fraser, A.F. and Terhune, M. (1977b) Radiographic studies of postural behaviour in the sheep fetus. II. Complex fetal movements. *Appl. Anim. Ethol.*, *3*, 235–246.

Fraser, D., Ritchie, J.S.D. and Fraser, A.F. (1975) The term 'stress' in a veterinary context. *Br. vet. J.*, *131*, 653–662.

Gary, L.A., Sherritt, G.W. and Hale, E.B. (1970) Behavior of Charolais cattle on pasture. *J. Anim. Sci.*, *30*, 203–206.

Gonyou, H.W., Christopherson, R.J. and Young, B.A. (1979) Effects of cold temperature and winter conditions on some aspects of behaviour of feedlot cattle. *Appl. Anim. Ethol.*, 5, 113–124.

Grandin, T. (1978) Design of lairage, yard and race systems for handling cattle in abbatoirs, auctions, ranches, restraining chutes and dipping vats (Diseño de sistemas de abrigos, naves y parques para el manejo de vacuno en mataderos, exposiciones, ranchos, mangas y baños para elganado). *Proc. 1st Wld congr. Ethology Appl. Zootechnics*, *1*, 37–52.

Grassia, A. (1978) A technique in the study of aggregative behaviour of sheep. *Appl. Anim. Ethol.*, *4*, 369–378.

Grubb, P. (1974) Mating activity and the social significance of rams in a feral sheep community. In *The Behaviour of Ungulates and Its Relation to Management*, ed. V. Geist and F. Walther, pp. 457–476. Morges, Switzerland: International Union for Conservation of Nature and Natural Resources.

Guhl, A.M. and Allee, W.C. (1944) Some measurable effects of social organization in flocks of hens. *Physiol. Zool.*, *17*, 320–347.

Hafez, E.S.E., Cairns, R.B., Hulet, C.V. and Scott, J.P. (1969) The behaviour of sheep and goats. In *The Behaviour of Domestic Animals*, ed. E.S.E. Hafez, 2nd ed., pp. 296–348. London: Baillière Tindall.

Hafez, E.S.E, Schein, M.W. and Ewbank, R. (1969) The behaviour of cattle. In *The Behaviour of Domestic Animals*, ed. E.S.E. Hafez, 2nd ed., pp. 296–348. London: Baillière Tindall.

Hafez, E.S.E. and Signoret, J.P. (1969) The behaviour of swine. In *The Behaviour of Domestic Animals*, ed. E.S.E. Hafez, 2nd ed., pp. 349–390. London: Baillière Tindall.

Hale, E.B. (1969) Domestication and evolution of behaviour. In *The Behaviour of Domestic Animals*, ed. E.S.E. Hafez, 2nd ed., pp. 22–42. London: Baillière Tindall.

Hartsock, T.G. and Graves, H.B. (1976) Neonatal behaviour and nutrition-related mortality in domestic swine. *J. Anim. Sci.*, *42*, 235–241.

Hemsworth, P.H., Beilharz, R.G. and Brown, W.J. (1978) The importance of the courting behaviour of the boar on the success of natural and artificial matings. *Appl. Anim. Ethol.*, *4*, 341–348.

Hemsworth, P.H., Winfield, C.G. and Mullaney, P.D. (1976) A study of the development of the teat order of piglets. *Appl. Anim. Ethol.*, *2*, 225–233.

Hergenhahn, B.R. (1976) *An Introduction to Theories of Learning*. New York: Prentice-Hall.

Heymer, A. (1977) *Ethological Dictionary*. Berlin and Hamburg: Paul Parey.

Hoffman, M.P. and Self, H.L. (1973) Behavioral traits of feedlot steers in Iowa. *J. Anim. Sci., 37*, 1438–1445.

Holmes, R.J. (1978) Veterinary and husbandry implications of the behaviour of prolific sheep lambing indoors (Les implications vétérinaires et d'élevage du comportement des brebis de races prolifiques lors de l'agnelage à l'intérieur). *Proc. 1st Wld Congr. Ethology appl. Zootechnics, 1*, 103–105.

Houpt, K.A. (1977) Horse behavior: its relevance to the equine practitioner. *Equ. Med. Surg., 1*, 87–94.

Houpt, K.A., Law, K. and Martinisi, V. (1978) Dominance hierarchies in domestic horses. *Appl. Anim. Ethol., 4*, 273–283.

Hughes, B.O. (1979) Aggressive behaviour and its relation to oviposition in the domestic fowl. *Appl. Anim. Ethol., 5*, 85–93.

Hughes, B.O. and Black, A.J. (1978) Agonistic behaviour in domestic fowl transferred between cages and pens. *Appl. Anim. Ethol., 4*, 181–186.

Hunter, R.F. and Milner, C. (1963) The behaviour of individual related and groups of south country Cheviot hill sheep. *Anim. Behav., 11*, 507–513.

Hurnik, J.F. (1978) Observational techniques in behavioural studies of farm animals (Technische Beobachtungsmethoden in der Verhaltenforschung bei landwirtschaftlichen Nutztieren). *Proc. 1st Wld Congr. Ethology appl. Zootechnics, 1*, 59–65.

Ingram, D.L. and Legge, K.F (1970) The thermoregulatory behaviour of young pigs in a natural environment. *Physiol. Behav., 5*, 931–987.

Keogh, R.G. (1978) Feeding behaviour and management of sheep and cattle (Conducta alimentaria y manejo de ovinos y bovinos). *Proc. 1st Wld Congr. Ethology appl. Zootechnics, 1*, 293–301.

Kiley-Worthington, M. (1977) *Behavioural Problems of Farm Animals*. Newcastle-upon-Tyne: Oriel Press.

Kiley-Worthington, M. and Savage, P. (1978) Learning in dairy cattle using a device for economical management of behaviour. *Appl. Anim. Ethol., 4*, 119–124.

Kilgour, R. (1969) *Social Behaviour in the Dairy Herd*, Publ. 459. Ruaraka: New Zealand Department of Agriculture.

Kilgour, R. (1978) Minimising stress on animals during handling (Réduction du choc ressenti par les animaux à la suite de manipulations). *Proc. 1st Wld Congr. Ethology appl. Zootechnics*, *1*, 303–322.

Kratzer, D.D. (1971) Learning in farm animals. *J. Anim. Sci.*, *31*, 1268–1273.

Lindsay, D.R. (1978) Effect of stimulation by partners on reproduction success (Effet de la stimulation par des partners sur le succès de la reproduction). *Proc. 1st Wld Congr. Ethology appl. Zootechnics*, *1*, 513–523.

Lindsay, D.R. and Fletcher, I.C. (1968) Sensory involvement in the recognition of lambs by their dams. *Anim. Behav.*, *16*, 415–417.

Lynch, J.J. and Alexander, G. (1976) The effect of gramineous windbreaks on behaviour and lamb mortality amongst shorn and unshorn Merino sheep during lambing. *Appl. Anim. Ethol.*, *2*, 305–325.

Lynch, J.J. and Alexander, G. (1977) Sheltering behaviour of lambing Merino sheep in relation to grass hedges and artificial windbreaks. *Aust. J. agric. Res.*, *28*, 691–701.

McBride, G. (1963) The 'teat order' and communication in young pigs. *Anim. Behav.*, *11*, 53–56.

McBride, G., Arnold, G.W., Alexander, G. and Lynch, J.J. (1967) Ecological aspects of the behaviour of domestic animals. *Proc. ecol. Soc. Aust.*, *2*, 133–165.

McBride, G., James, J.W. and Shoffner, R.N. (1963) Social forces determining spacing and head orientation in a flock of domestic hens. *Nature, Lond.*, *197*, 1272–1273.

McNeilly, A.S. and Ducker, H.A. (1972) Blood levels of oxytocin in the female goat during coitus and in response to stimuli associated with mating. *J. Endocrin.*, *54*, 399–406.

Malechek, J.C. and Smith, B.M. (1976) Behaviour of range cows in response to winter weather. *J. Range Mgmt*, *29*, 9–11.

Mimura, K. (1978) Feeding behaviour of cattle under warm and humid climate (Comportement alimentaire du bétail sous climat chaleureux et humide). *Proc. 1st Wld Congr. Ethology appl. Zootechnics*, *1*, 417–421.

Morgan, P.D. and Arnold, G.W. (1974) Behavioural relationships between Merino ewes and lambs during the four weeks after birth. *Anim. Prod.*, *19*, 169–176.

Morgan, P.D., Boundy, C.A.P., Arnold, G.W. and Lindsay, D.R. (1975) The roles played by the senses of the ewe in the location and recognition of lambs. *Appl. Anim. Ethol.*, *1*, 139–150.

Moss, B.W. (1978) Some observations on the activity and aggressive behaviour of pigs when penned prior to slaughter. *Appl. Anim. Ethol.*, *4*, 323–339.

Mueller, C.G. and Rudolph, M. (1966) *Light and Vision,* Life Science Library. New York: Time Life Books.

Naaktgeboren, C. and Slijper, E.J. (1970) *Biologie Der Geburt.* Hamburg and Berlin: Paul Parey.

Olds, J. (1977) *Drives and Reinforcements, Behavioural Studies of Hypothalamic Functions.* New York: Raven Press.

Perry, G.C., Patterson, R.C.S. and Stinson, G.C. (1972) Submaxillary salivary gland involvement in porcine mating behaviour, VII. *Int. Konf. tierische Forpplanzunof.*, *1*, 395–399.

Reinhardt, V., Mutiso, F.M. and Reinhardt, A. (1978*a*) Social behaviour and social relationships between female and male prepubertal bovine calves (*Bos indicus*). *Appl. Anim. Ethol.*, *4*, 43–54.

Reinhardt, V., Mutiso, F.M. and Reinhardt, A. (1978*b*) Resting habits of Zebu cattle in a nocturnal enclosure. *Appl. Anim. Ethol.*, *4*, 261–271.

Reinhardt, V. and Reinhardt, A. (1975) Dynamics of social hierarchy in a dairy herd. *Z. Tierpsychol.*, *38*, 315–323.

Rossdale, P.D. (1970) Perinatal behaviour in the Thoroughbred horse. *Br. vet. J.*, *126*, 656.

Rossdale, P. and Ricketts, S. W. (1974) *The Practice of Equine Stud Medicine.* London: Baillière Tindall.

Ruckebusch, Y. (1975) The hypnogram as an index of adaptation of farm animals to changes in their environment. *Appl. Anim. Ethol.*, *2*, 3–18.

Ruckebusch, Y. and Bueno, L. (1978) An analysis of ingestive behaviour and activity of cattle under field conditions. *Appl. Anim. Ethol.*, *4*, 301–313.

Sambraus, H.H. (1979) A review of historically significant publications from German speaking countries concerning the behaviour of domestic farm animals. *Appl. Anim. Ethol.*, 5, 5–13.

Schäfer, M. (1975) *The Language of the Horse.* London: Kay and Ward (New York: Arco Publishing).

Schjelderup-Ebbe, T. (1931) Die Despotie in sozialen Laben der Vogel. *Forsch. Volkerpsychol. sozialog.*, *10*, 77–140.

Scott, J.P. (1945) Social behavior, organization and leadership in a small flock of domestic sheep. *Comp. Psychol. Monogr.*, *18*, 1–29.

Sharafelden, M.A. and Shafie, M.M. (1965) Animal behaviour in Subtropics. Part II. Grazing behaviour of sheep. *Netherlands J. agric. Sci.*, *13*, 239–247.

Sherritt, G.W., Graves, H.B., Gobble, J.L. and Hazlett, V.E. (1974) Effects of mixing pigs during the growing–finishing period. *J. Anim. Sci.*, *39*, 834–837.

Sherritt, G.W., Orr, D.E., Jr., Gobble, J.L., Hazlett, V.E., Aldrich, R.A. and Partenheimer, E.J. (1972) A swine production system based on the use of one pen from birth to market. *J. Anim. Sci.*, *34*, 709–712.

Shillito-Walser, E. (1978) Maternal behaviour and management to minimise postnatal deaths (Das Verhalten der Mutter zu ihren Junger). *Proc. 1st Wld Congr. Ethology appl. Zootechnics*, *1*, 285–292.

Shillito, E.E. and Alexander, G. (1975) Mutual recognition amongst ewes and lambs of four breeds of sheep (*Ovis aries*). *Appl. Anim. Ethol.*, *1*, 151–165.

Signoret, J.P. (1970) Effect of disease and stress on reproductive efficiency in swine. In *Swine Behaviour and Reproduction*. Symposium Proceedings 70, pp. 28–45. Lincoln, Ne.: University of Nebraska College of Agriculture.

Squires, R.V. (1978) Effects of management and the environment on animal dispersion under free range conditions (L'influence de l'aménagement et du milieu sur la dispersion du bétail en pâturage libre). *Proc. 1st Wld Congr. Ethology appl. Zootechnics. 1*, 323–334.

Stephens, D.B. (1974) Studies on the effect of social environment on the behaviour and growth rates of artificially reared male calves. *Anim. Prod.*, *18*, 23–24.

Stephens, D.B. and Linzell, J.L. (1974) The development of sucking behaviour in the new-born goat. *Anim. Behav.*, *22*, 628–633.

Stricklin, W.R., Wilson, L.L. and Graves, H.B. (1976) Feeding behaviour of Angus and Charolais–Angus cows during summer and winter. *J. Anim. Sci.*, *43*, 721–732.

Syme, G.H., Syme, L.A. and Jefferson, T.P. (1974) A note on the variations in the level of aggression within a herd of goats. *Anim. Prod.*, *15*, 309–312.

Thiessen, D.D. (1964) Population density and behaviour: a review of theoretical and physical contributions. *Tex. Rep. Biol. Med.*, *22*, 266–314.

Thinès, G., Soffié, M. and de Marneffe, G. (1975) Aires de résidence préférentielles d'un groupe de vaches laitières en stabulation libre. *Ann. Zootech.*, *24*, 177–187.

Tindell, R. and Craig, J.V. (1959) Effects of social competition on laying house performance in the chicken. *Poultry Sci.*, *38*, 95–105.

Tolman, G.W. (1965) Emotional behaviour and social facilitation of feeding in domestic chicks. *Anim. Behav.*, *13*, 493–496.

Tolman, G.W. (1968) The role of the companion in social facilitation of animal behaviour. In *Social Facilitation and Imitative Behaviour*, ed. Simmiel, Hoppe and Milton. Algar and Barcon.

Torres-Hernandez, G. and Hohenboken, W. (1979) An attempt to assess traits of emotionality in crossbred ewes. *Appl. Anim. Ethol.*, *5*, 71–83.

Tribe, D.E. (1950) The behaviour of the grazing animal. A critical review of present knowledge. *J. Br. Grassl. Soc.*, *5*, 209–224.

Tribe, D.E. and Gordon, J.G. (1949) The importance of colour vision to the grazing sheep. *J. agric. Sci., Camb.*, *39*, 313–314.

Van der Lee-Boot, C.M. (1956) Spontaneous pseudo-pregnancy in mice, II. *Acta physiol. pharmac. Neer.*, *5*, 213.

Van Putten, G. (1969) An investigation into tail biting among fattening pigs. *Br. vet. J.*, *125*, 511.

Verplanck, W.S. (1957) A glossary of some terms used in the objective science of behavior. *Psychol. Rev.*, Suppl. *64*, 1–42.

Wagnon, K.A. (1965) Social dominance in range cows and its effect on supplemental feeding. *Div. agric. Sci. Univ. Calif. Bull.*, 819.

Waring, G.H., Wierzbowski, S. and Hafez, E.S.E. (1975) The behaviour of horses. In: *The Behaviour of Domestic Animals*, ed. E.S.E. Hafez, 3rd ed., pp. 330–369. London: Baillière Tindall.

Welch, R.A.S. and Kilgour, R. (1970) Mismothering amongst Romneys. *N.Z.J. Agric.*, *121*, 26–27.

Whitten, W.K. (1956) Modification of mouse oestrous cycle by external stimuli associated with the male. *J. Endocr.*, *13*, 399–404.

Whittlestone, W.G., Albright, J.L. and Kilgour, R. (1972) Learning and behaviour patterns in cows. *Aust. Dairy J.*, *65*, 46–47 (abstr.).

Wierzbowski, S. (1959) The sexual reflexes of stallions. *Rocz. Nauk. Roln.*, *73-B-4*, 753–788.

Williamson, N.B., Morris, R.S., Blood, D.C., Cannon, C.M. and Wright, P.J. (1972) A study of oestrous behaviour and oestrus detection methods in a large commercial dairy herd. 2. Oestrous signs and behaviour patterns. *Vet. Rec.*, *91*, 58–61.

Wilson, E.O. (1975) *Sociobiology: The New Synthesis*. Harvard, Mass.: Belknap Press of Harvard University Press.

Wilson, R.K. and Flynn, A.V. (1974) Observations on the eating behaviour of individually fed beef cattle offered grass silage ad libitum. *Irish J. agric. Res.*, *13*, 347.

Wilson, R.K. and Flynn, A.V. (1975) A note on the eating behaviour of cattle offered grass silage ad lib in troughs. *Irish J. agric. Res.*, *14*, 218–220.

Wilson, R.K. and Flynn, A.V. (1976) The eating behaviour of steers offered grass silage ad libitum in troughs with and without a barley supplement. *Proc. Nutr. Soc.*, *35*, 15A.

Wood, P.D.P., Smith, G.F. and Lisle, M.F. (1967) A survey of intersucking in dairy herds in England and Wales. *Vet. Rec.*, *81*, 396–398.

Wood-Gush, D.G.M. (1971) *The Behaviour of the Domestic Fowl*. London: Heinemann.

Zeeb, K. (1978) Klimafaktoren-Einfluss aud die Aktivität von Rindern in verschiedenen Haltungssystemen (Effects on cattle of climatic factors under various systems of management). *Proc. 1st Wld Congr. Ethology appl. Zootechnics*, *1*, 115–128.

Zito, C.A., Wilson, L.L. and Graves, H.B. (1977) Some effects of social deprivation on behavioral development of lambs. *Appl. Anim. Ethol.*, *3*, 367–377.

Glossary of Terms in Applied and Veterinary Ethology *

Aerophagia: Pathological and excessive swallowing of air: a vice in horses.

Aggressive Behaviour: The tendency to initiate a vigorous conflict.

Agonistic Behaviour: Any behaviour associated with conflict or fighting between two individuals. It includes patterns of behaviour involving escape or passivity.

Allelomimetic Behaviour: Any behaviour where animals perform the same activity with some degree of mutual stimulation and consequent co-ordination.

Anomalous Behaviour: Behaviour which is a variant of a normal activity but displayed abnormally, e.g. excessive self-grooming or locomotor stereotypies.

Anorexia: Abnormal lack of ingestive behaviour, e.g. in toxic and depressed clinical states.

Aversion Therapy: Treatment of a compulsive form of behaviour by associating the behaviour with an electric shock, or other aversive stimulation.

Competition: (1) The direct struggle between individuals for a limited supply of environmental necessities. (2) The common striving for living requirements such as food, space or shelter, by two or more individuals, populations or species.

---

* For further reading, see
Heymer, A. (1977) *Ethological Dictionary*. Berlin and Hamburg: Paul Parey.
Wilson, E.O. (1975) Glossary. *Sociobiology*, pp. 578–598. Harvard, Mass.: Belknap Press of Harvard University Press.

Conditioning: (1) This occurs when a reflex is modified by specific experience. (2) It occurs in a conditioned reflex when the original stimulus has been substituted. (3) The process by which an animal acquires the capacity to respond to a given stimulus with the reflex reaction proper to another stimulus (the re-inforcement) when the two stimuli are applied concurrently a number of times.

Consummatory Act: An act which constitutes the termination of a given instinctive behaviour pattern.

Contactual Behaviour: Maintenance of bodily contact. The formation of simple aggregations through behaviour of this sort occurs very commonly in animals.

Critical Period: The infantile and maternal phases when the subject is most sensitive to specific environmental features and experiences.

Displacement Activity: (1) An activity belonging to an instinct other than the one activated. (2) An activity performed by an animal in which two or more compatible drives are strongly activated. (3) Activities performed by an animal in which one drive is, at the same time, both activated and thwarted. (4) The performance of a behaviour pattern out of the context of behaviour to which it is normally related.

Dominance: (1) An individual animal is said to be dominant over another when it has priority in feeding and sexual behaviour, and when it is superior in aggressiveness and in group control. (2) Dominance status is indicated by superiority in fighting ability of one individual over one or more species mates. (3) A dominant animal is one the behaviour of which proceeds without reference to the behaviour patterns of a subordinate animal.

Drive: The complex of internal and external states and stimuli leading to a given behaviour pattern.

Eliminative Behaviour: Patterns of behaviour connected with evacuation of faeces and urine.

Epimeletic Behaviour: The provision in behavioural terms of care or attention; includes suckling in particular.

Et-epimeletic Behaviour: Care-seeking behaviour in the young animal; includes soliciting maternal attendance in particular.

Ethogram: An inventory of behaviour patterns typical of an animal or species.

Ethostasis: The repression, through environmental controls, of major items of instinctive behaviour in livestock.

Flight Reaction: A characteristic escape reaction, specific for a particular enemy and surroundings, occurring as soon as the intruder approaches.

Flight Distance: That radius of surrounding areas within which intrusion provokes a flight reaction.

Habituation: The permanent weakening of a response as a result of repeated stimulation unaccompanied by re-inforcement. This is regarded as distinct from fatigue.

Hierarchy: Any social rank or order established through direct combat, threat, passive submission or some combination of these behaviour patterns.

Home Range: That locality where the individual animal imprints its principal functions.

Imprinting: (1) A rapid and usually very stable form of learning taking place in early life. (2) The infantile parameter whereby, often without any apparent immediate re-inforcement, broad supra-individual characteristics of the species come to be recognized as the species pattern and subsequently used as releasers.

Ingestive Behaviour: Behaviour concerned with the selection and consumption of food and drink.

Innate Releasing Mechanism: A hypothetical mechanism which is to prevent all discharge of activity unless the animal encounters the right environmental situation to release this block.

Intersucking: Abnormal sucking activity directed to appendages of others in groups of young livestock prematurely weaned.

Instinct: (1) An inherited and adapted system of coordination within the nervous system as a whole which, when activated, finds expression in behaviour culminating in a fixed pattern. (2) A general term applied to behaviour differences which are largely determined by heredity.

Kinesis: An undirected reaction, without orientation of the body in relation to the stimulus.

Leadership: A special form of facilitation, in which one animal sets the pace of group activity or initiates changes in it.

Learning: The process which produces adaptive change in individual behaviour as the result of experience. It is regarded as distinct from fatigue, sensory adaptation, maturation and the consequences of bodily changes of any kind.

Libido: Used synonymously with male sex drive in domesticated animals.

Motive State (Motivity): The behavioural manifestations of a given state of motivation, e.g. threat postures indicating the physiological condition of fight-or-flight.

Overflow Activities: Reactions to meagre or abnormal stimuli.

Peck Dominance: The dominance of one individual over others in most of their contacts.

Peck Order: (1) The rank of several members within a social hierarchy. (2) Arrangements according to dominance.

Pheromone: (1) A substance secreted by one individual and received by a second individual of the same species, releasing a specific reaction of behaviour or a developmental process. (2) Ectohormone.

Phonation: Expression by sound.

Polydipsia Nervosa: Excessive drinking of water beyond physiological needs, e.g. excess water intake in horses as a stable vice.

Reaction Time: Time between application of stimulus and response of whole animal.

Reflex: An innate and simple response involving the central nervous system and occurring very shortly after the stimulus which evokes it. It characteristically involves only a part of the organism, though the whole may be affected, and is usually a response to localized stimuli.

Releasers: Characters which are peculiar to individuals of a given species and to which responsive releasing mechanisms of other individuals react, thus setting in motion chains of instinctive actions.

Ritual Behaviour: (1) Behaviour which has lost its original meaning and acquired a new one as a means of expression. (2) An originally variable sequence of behavioural actions which has become an unchangeable ceremony.

Social Behaviour: (1) The reciprocal interactions of two or more animals and the resulting modifications of individual action systems. (2) Any behaviour caused by or affecting another animal, usually one of the same species.

Social Facilitation: Synchrony of certain activities and some drive increases under the influence of group effects.

Social Organization: An aggregation of individuals into a fairly well-integrated and self-consistent group in which the unity is based upon the interdependence of the separate organisms.

Social Releaser: Any specific or complex feature of an organism eliciting an instinctive activity in another individual of the same or another species.

Stimulus, Primary or Releaser: Specific stimulus to which certain responses are automatically given. Such responses form the basis of learning, in early development.

Taxis: (1) Locomotion either directly towards or away from a source of stimulation. (2) Locomotory behaviour involving a steering reaction. (3) The spatial correction movement resulting in orientation.

Territoriality: Proprietary behaviour in respect of defence of all or part of the home range of an animal. This defence is directed primarily against members of the same species.

Thigmotaxis/Thigmotropism: Behaviour revealing a drive to make and maintain close bodily contact with an associate animal.

Tonic Immobility: (1) A state of locomotor economy, shown particularly in an unwillingness to make responses which involve complex, co-ordinated bodily movements. (2) An apparent absence of co-ordinated responses in an animal without an associated physical impairment.

Index

A

Abnormal behaviour, 79, 231–64
  anomalous, 240–56
  clinical disorders, 257–64
  maternal, 227–8
  stress and, 233–9
  sexual, 248–50
Abscess, spinal, 260–1
Acetylcholine, 38, 144
Acquired behaviour, 36–46
Acquisitive reactions, 112
Activators, 12
Acute stress syndrome, 235
Adaptive responses, 233
  postnatal, 66
  and posture, 260–1
Adrenal glands, 16–17
Adrenaline, 16–17, 234
Aerophagia, 246, 275
Aggressive behaviour, 89, 96, 114, 275
  cattle, 169
  horses, 151
  mechanisms, 22–3, 26
  pigs, 183
  and overcrowding, 242
  and territorialism, 136, 140
  towards young, 227–8, 251
Agonistic behaviour, 16, 91, 92, 96,
    124, 179, 188, 275
Alarm reactions, 17, 38
Aldosterone, 104
Allelomimesis, 41, 53, 177, 275
Ambulation, postnatal, 67
Androgenized females, 96
Androgens, 16, 34
Anestrus, 208
Angiotensin, 104
Anomalous behaviour, 114–15, 240–
    56, 275
Anorexia, 275
Antidiuretic hormone, 104
Appetite
  depraved, 108, 246, 248
  specific, 103–4
  see also Food selectivity and Ingestion
Arthritis, 260

Association, 26, 53, 123–7, 278
  cattle, 166–8
  horses, 154–6
  pigs, 182–4
  poultry, 190–2
  sheep, 177–8
Auditory area of cerebral cortex, 13
Auditory stimuli, 20
Aversion therapy, 275
Avoidance, 39

B

Bark-eating, 108
Batchelor groups, 124, 198
Behaviour
  abnormalities, 231–64
  acquired, 36–46
  anomalous, 108, 240–56
  association, 26, 53, 123–7, 154–6,
    166–8, 177–8, 182–4, 190–2,
    278
  body care, 90, 128–33, 156–8, 168–
    9, 178, 184–6, 192–3
  cattle, 78–81, 160–70
  cerebellar, 26
  and clinical disorders, 257–64
  cord, 26, 92
  cortical, 26
  developmental, 57–85
  diagnostic aid, 258–63
  diencephalic, 26
  environmental influences, 57–56
  exploratory, 68, 89–90, 109–15,
    153, 165, 177, 181, 190
  fetal, 59–65
  group, 24–5, 53–4, 107
  horses, 71–8, 151–9
  ingestion, 24, 25, 48–9, 100–8, 153,
    161–5, 172–7, 179–80, 189–90
  kinetic, 59–65, 90, 116–22, 153–4,
    166, 177, 181–2, 190
  maintenance, 81, 87–148
  maternal, 24, 43, 224–30, 251–2
  motivation and, 21–5
  nature, 11–56

Behaviour (*continued*)
  nursing, 224–30
  organization of reactions, 26–30
  parturient, 212–23
  pigs, 179–94
  postnatal, 66–70
  poultry, 84–5, 188–94
  reactivity, 92–9
  ready state, 11–20
  reproductive, 49, 195–230
  rest and sleep, 75, 83, 91, 143–8,
    158–9, 170, 178, 187, 193–4
  sexual, 24, 34–5, 49, 197–211,
    248–50
  sheep and goats, 81, 171–8
  species patterns, 149–94
  stress and, 233–9
  territorial, 91, 124–42, 158, 169–70,
    186–7, 193, 202, 279
'Biological clock', 143
Biostimulation, 207
Birth, 212–23
Bloat, 261
Body care, 90, 128–33
  cattle, 168–9
  horses, 156–8
  pigs, 184–6
  poultry, 192–3
  sheep and goats, 178
Body sense, 13
Body sleep, 144
'Boredom', 237–8
Brain hormones, 18
Brain sleep, 143
Brambell Report, 5–7
Browsing, 108
Butting, 160, 171

C

Cannibalism, 222, 227–8, 244, 246,
  251
Cannon, W. B., 234
Canter, 119, 154
Captivity, 237
Care-seeking, 226–7

Catecholamines, 23, 29
Cattle
  anomalous behaviour, 244–5
  association, 166–8
  body care, 168–9
  coitus, 210
  development, 78–81
  estrus, 205
  exploration, 165
  ingestion, 48–9, 161–5
  kinesis, 166
  parturition, 218–19
  reactivity, 160–1
  rest and sleep, 146–8, 170
  restraint, 54–6
  territorialism, 169–70
  thermoregulation, 47–8
Cerebellar behaviour, 26
Cerebral cortex, 12–14, 44–5
Challenging behaviour, 201
Clasping during coitus, 211
Climate and behaviour, 47–52, 105
Clinical disease and behaviour, 257–
  64
Cloaca, 192
Coat condition in illness, 130
Cock-crowing, 193
Coital behaviour, 209–11
Colic, 263
Coliform enteritis and stress, 235
Colostrum, 228
Colour vision, 32, 190
Comfort-seeking, 128, 130, 192
Competition, 275
Complex hierarchies, 167
Conditioning, 38–9, 276
Confinement, 109, 140
  see also Restraint
Consummatory act, 276
Contactual behaviour, 276
Copulation, 208–11
Cord behaviour, 26, 92
Cortical behaviour, 26
Cows, *see* Cattle
Crib-biting, 246
Critical period, *see* Imprinting
Crowding, *see* Stocking density

Crushes, 54–5
Cud-chewing, *see* Rumination

D

Darwin, C., 234
Day length and behaviour, 19, 51, 144
Debeaking, 127, 191, 246
Defaecation, *see* Elimination
Developmental behaviour, 57–85
    calves, 78–81
    fetal, 59–65
    foals, 71–8
    lambs, 81
    piglets, 81–4
    poultry, 84–5
    postnatal, 66–70
Diarrhoea and stress, 235
Diencephalic behaviour, 26
Digital pads, fetal, 68
Displacement activity, 276
Diurnal rhythms, 143, 144
    see also Photoperiodism
DNA, 28, 29, 143
'Dog-sitting', 260
Dominance, 26, 96, 124–6, 136, 139,
    155, 166–8, 183, 276, 278
Dopamine, 23
Dreaming, 144
Drive, 23–5, 276
Dust-bathing, 192–3

E

EAM phenomenon, 31
Eating, *see* Food selection *and*
    Ingestion
EEG, sleeping, 144
Elimination, 27, 82, 90, 92, 130, 141,
    276
    cattle, 168–9
    horses, 157–8
    pigs, 186
    poultry, 192
Emotionality, 22–3

Empiricism in behaviour, 39, 110
Endocrine system, 14–18
Energy intake, 101
Enteritis and stress, 235
Environment
    and behaviour, 47–56
    and estrus, 204–5
    quality and quantity, 240–1
    and stress, 238
Epimeletic behaviour, 276
Eponychia, 68
Eructation reflex, depressed, 261
Estrogen, 17, 34
Estrus behaviour, 17, 94–6, 98, 204–8
    anomalies, 250
Et-epimeletic behaviour, 226–7, 276
Ethogram, 276
Ethology, 1–7
'Ethostasis', 28
Examination of animal groups, 259
Exploration, 89–90, 109–15
    cattle, 165
    horses, 153
    pigs, 181
    postnatal, 68
    poultry, 190
    sheep, 177
Exteroceptors, 11
Eye, 31, 94

F

Facial expression and pain, 262
'False mounting', 209
FAP, 30, 169
Feather-picking, 127, 191, 246
Feeding, *see* Ingestion
Female sexual behaviour, 204–8
    anomalies, 250
Fetal behaviour, 59–65
Fight-or-flight reactions, 16, 23, 91,
    92, 138
Fighting
    boars, 179
    cattle, 160–1
    poultry, 188, 192
    rams, 170

Fixed action patterns, 30, 169
Flehmen, 33, 198, 199, 217
Flight distance, 135, 137, 277
Flight reaction, 16, 276
   see also Fight-or-flight reaction
Fly problems, 128
Follicle-stimulating hormone, 15
Following, *see* Leader–follower relationships
Food intake, *see* Ingestion
Food selection, 162–3, 174, 180, 190
Foot pain, 261
Fostering, 79
Fractures, 260
von Frisch, K., 1

G

Gait, 117–20, 153–4
Gallop, 119, 154
GAS, 234–5
Gastric ulcers and stress, 235
'Gating', 254
General adaptation syndrome, 234–5
Glands, *see specific gland*
Goats, *see* Sheep and goats
Gonadotropic hormones, 15
Gonitis, 260
Grazing, 105–7, 153, 161–3, 172–5
   and kinesis, 117
   postnatal, 43, 74–5
   seasonal changes, 99, 171
Grooming, 90, 131, 156–7, 168, 192
   excessive self, 130
   inhibited, 261
   neonatal, 76, 215, 225
      refusal, 251, 252
   see also Body care
Group behaviour, 24–5, 53–4, 107
Grunting, 152, 179
Gustatory stimuli, 20

H

Habituation, 41, 277
Hair balls, 131

Handling, 39, 54–6
Hatching, 84, 127
Head-shaking, 248
Head space, 135, 137
Hearing, 20
Heat, *see* Estrus
Hierarchies, *see* Social hierarchies
Home range, 126, 134, 140, 153, 277
Homosexual activity in rams, 245
Hormones, 14–18, 33–5, 144
Horn-rubbing, 131
Horses
   anomalous behaviour, 246–8
   association, 154–6
   body care, 156–8
   coitus, 210
   development, 71–8
   estrus, 205
   exploration, 153
   ingestion, 153
   kinesis, 153–4
   parturition, 217–18
   reactivity, 151–2
   rest and sleep, 144–6, 158–9
   restraint, 55
   territorialism, 158
Housed feeding behaviour, 107
Huddling, 133, 185
Humerus, fracture, 260
Hunger, 24, 98, 100
   see also Ingestion
Husbandry systems and stress, 240–1
Hygiene, *see* Body care
Hyperactivity, 92
Hypnograms, 148
Hypokinaesthesia, 254–6
Hypostimulation, 235, 254
Hypothalamus, 14, 22, 23, 45, 100, 101, 143
Hypotonia, 96–7, 255

I

Imitation, 41, 53, 177, 275
Immobility, 279
Impotence, 248–9

Imprinting, 41–3, 84, 126, 224, 276, 277
Infancy, acquired behaviour in, 36–8
Ingestion, 24, 25, 100–8, 277
  and climate, 48–9
  cattle, 161–5
  horses, 153
  pigs, 179–80
  postnatal, *see* Suckling
  poultry, 189–90
  sheep and goats, 172–7
  see also Food selection *and* Grazing
Innate releasing mechanisms, 277
Inquisitive reactions, 112
Instinct, 277
Intelligence, 45–6
Interoceptors, 12
Intersucking, 79, 126, 244, 277
Intromission, 209–11
Investigation, *see* Exploration
Isolation, 53, 82, 114

J

Jacobsen's organ, 33
'Judas' animals, 126
Jumping, 154

K

Kicking, 141
Kinesis, 90, 116–22, 227
  cattle, 166
  fetal, 59–65
  horses, 153–4
  pigs, 181–2
  poultry, 190
  sheep and goats, 177
  see also Play

L

Lactiferous reflex, 70
Lamb-stealing, 220, 251, 252

Laminitis, 260, 261
Leader–follower relationships, 26, 42, 127, 167–8, 277
Learning, 38–45, 110, 227
Libido, 49–50, 197–8, 277
  anomalies, 248–50
Light stimuli, *see* Visual stimuli *and* Photoperiodism
Limb reflexes, 92
Limbic system, 14, 22–3
Linear hierarchy, 167
Locomotion, 117–20, 153
  postnatal, 66
Lorenz, K. Z., 1–2, 30
Luteinizing hormone, 15

M

Maintenance behaviour, 81, 87–148
  association, 90, 123–7, 154–6, 166–8, 177–8, 182–4, 190–2
  body care, 90, 128–33, 156–8, 168–9, 178, 184–6, 192–3
  exploration, 89–90, 109–15, 153, 165, 177, 181, 190
  ingestion, 24, 100–8, 153, 161–5, 172–7, 179–80
  kinesis, 90, 116–22, 153–4, 166, 177, 181–2, 190
  reactivity, 92–9, 151–2, 160–1, 171–2, 179, 188–9
  rest and sleep, 91, 143–8, 158–9, 170, 178, 187, 193–4
  territorialism, 91, 134–42, 158, 169–70, 178, 186–7, 193
Male sexual behaviour, 197–204
  anomalous, 248–50
Maternal behaviour, 24, 43, 224–8
  anomalous, 251–2
Maternal desertion, 252
Mating behaviour, 208–11
Mating inexperience, 249
Mechanical conditions and posture, 260
Melatonin, 18, 144
Membranes, delivery of, 215

Memory, 44–5
Metacarpus, fracture, 260
Milk let-down, 70
Milking and grazing, 107
Milking order, 126
Mineral deficiency, 104, 108
Motivation, 21–2, 277
Mounting, 209, 249
Mouthing, 243
Muskone, 207

N

Nasal secretion, 132, 261
Negative reinforcement, 39
Neighing, 152
Neonatal behaviour, 66–85, 228–30
    calves, 78–81
    foals, 71–8
    lambs and kids, 81
    piglets, 81–4
    poultry, 84–5
Nerve impulses, 11
Nervous conditions and posture, 260
Nervous system, 11–14, 27–30
Nest-building, 213, 221
Neurons, 12, 28
Neurophysiology, 28–30
'Nipping', 131
Nissl substance, 28, 29
Noesis, 45–6
Norepinephrine, 29
'Nudging', 203–4
Nursing, see Suckling
Nursling hairs, 69

O

Oestrogen, see Estrogen
Oestrus, see Estrus
Observation, learning by, 43
Olfaction, 18–19, 32–3
Olfactory area of cerebral cortex, 13
Olfactory reflex, 33, 198, 199, 217
Operant conditioning, 39, 46

Oral activity, pathological, 252–4
Orientation, 92, 128
    postnatal, 68–9
Osteomyelitis, 260, 261
Ovary, 17
Overcrowding, see Stocking density
Overflow activity, 278
Ovulation, 17
Oxygen consumption and kinesis, 121
Oxytocin, 16

P

Pain
    before and during birth, 213–14
    as diagnostic aid, 262–3
    feet, 261
    manifestations of, 262–3
    and postures, 260
Pair bonds, 123
Paradoxical sleep, 144
Parturient behaviour, 212–23
Pathological oral activity, 252–4
Pavlov, 38
Peck order, 127, 191, 278
Peptides, 23
Perceptive needs, 114
Personal space, 135, 137, 140
Pheromones, 32–3, 207, 278
Phonation, 278
Photoperiodism, 19, 51, 99, 144
    see also Day-length and Seasonal
        breeding
Phosphorus deficiency, 108
Physical space, 135, 136
Physiological changes under stress,
    234
Pica, 108, 246, 248
Pigs
    anomalous behaviour, 242–4
    association, 182–4
    body care, 184–6
    coitus, 210
    development, 81–4
    estrus, 206
    exploration, 181

Pigs (*continued*)
  handling, 56
  ingestion, 181
  kinesis, 181–2
  parturition, 221–2
  reactivity, 179
  rest and sleep, 187
  sexual behaviour, 49
  territorialism, 186–7
Pineal gland, 18, 144
Pituitary gland, 15–16
Placenta, delivery of, 215
Placentophagia, 215–16, 219
Play, 70, 79–80, 81, 83, 166, 181
Polydipsia nervosa, 105, 248, 278
Positive reinforcement, 39
Postnatal behaviour, 66–70
Posture
  as diagnostic aid, 259–60
  defects, and fetal inertia, 62–3
  during birth, 215, 217, 219, 220, 222
  elimination, 130, 157, 169, 192
  estrus, 205, 206
  neonatal, 228
  and pain, 263
  rest and sleep, 143, 145, 147, 148, 158, 170, 193
  submissive, 188
  threat, 96–7, 139, 160, 169, 200
Poultry
  anomalous behaviour, 245–6
  association, 127, 190–2
  body care, 192–3
  central nervous system, 27
  development, 84–5
  exploration, 190
  ingestion, 189–90
  kinesis, 190
  reactivity, 188–9
  rest and sleep, 193–4
  territorialism, 193
Preening, 192
Pre-parturient behaviour, 212–13
Primary releaser, 278
Productivity and stress, 238
Protein and memory, 44

Psychic impotence, 249
Punishments, 22, 41

R

Rabies, 115
Radial paralysis, 260
Rail transport, 52
Range animals and grazing, 106
Ready state, 11–20
Reaction time, 278
Reactions, 26–30
Reactivity, 89, 92–9
  cattle, 160–1
  horses, 151–2
  pigs, 179
  poultry, 188–9
  sheep and goats, 171–2
Receptivity, *see* Estrus
Recumbency, coordinating postnatal, 66
Reflexes, 89, 92–4, 278
  as diagnostic aid, 261–2
  righting, 63
Reinforcement, 38–9
'Releasers', 277, 278
REM sleep, 143
Reproductive behaviour, 195–230
  and climate, 49
  nursing and maternal, 224–30
  parturient, 212–13
  sexual, 197–211
Respiratory action and kinesis, 120
Rest and sleep, 91, 143–8
  cattle, 170
  horses, 158–9
  pigs, 187
  postnatal, 75, 83
  poultry, 193–4
  sheep and goats, 178
Restraint, 54–6, 130, 159, 237, 238, 243, 244
Retinex, 32
Rewards, 22
Righting reflexes, 63
Ritual behaviour, 278

RNA, 28, 29, 143
Rolling, 131, 157, 159
Roosting, 193–4
Rooting, 180
Rumination, 107, 163–4, 170, 175

S

Salt appetite, 103–5
Salt deficiency, 163
Satiety centre, 24, 101–3
Scotophobin, 44
Scours and stress, 235
Scrapie, 131
Scratching, 130, 131
Scrubbing of buttocks, 132
Sea transport, 53
Seasonal breeding, 19, 49–52, 99, 171, 197
Self-determination, 114
Selye, H., 234
Sense organs, 12
Sensitive period, see Critical period
Serotonin, 144
Sex hormones, 34
Sexual behaviour, 24, 34–5, 197–211
    and climate, 49
    anomalous, 248–50
    female, 204–8
    male, 197–204
    see also Breeding
Sexual organs, 17
Shade-seeking, 47–8, 128, 133
Shaking, 131
'Shaping', 39–41
Sheep and goats
    anomalous behaviour, 245
    association, 177–8
    body care, 178
    coitus, 210
    development, 81
    estrus, 205–6
    exploration, 177
    ingestion, 172–7
    kinesis, 177
    parturition, 220–1

Sheep and goats (continued)
    reactivity, 171–2
    rest and sleep, 178
    sexual behaviour, 49, 51
    territorialism, 178
Sight, see Vision
Skin, 33
Sleep, see Rest and sleep
Slow-wave sleep, 143
Smell, 18–19, 32–3
Social behaviour, see Association
Social density, see Stocking density
Social hierarchies, 90, 112, 123–6, 136, 139, 155, 166–8, 182, 191–2, 201, 242, 277
Social organization, 278
Social releaser, 278
Social space, 135
Socialization, 123
Sodium appetite, 103–5
Sodium deficiency, 163
Somasthetic area of cerebral cortex, 13
Somnolence in sexual behaviour, 250
'Sourness', 247
Space and behaviour, 134–42
Spastic paresis, 260, 261
Spinal abscess, 260–1
Squealing, 152
Stable vices, 246–8
Stall-kicking, 247
Stance, see Posture
Standing, postnatal, 66–7, 71, 78, 110, 228
Stereotypy, 28, 246–7
Stimulation, 11, 18–20, 114, 235, 278
    restricture, 254–6
Stocking density, 53, 124–5, 137, 140, 169, 186–7, 191, 193, 242, 245–6
Stocks, 54–5
Straining during birth, 214–15
Stress, 233–9
Stretching, 120–1
    inhibited, 261
Stride, 118
Subestrus, 250

Submissive response, 96–7
    see also Dominance
Subordinance, see Dominance
Suckling, 70, 74, 78–9, 82, 175–7,
        224–30
    refusal, 251, 252
Superdrives, 53
Sympathetic nervous system, 234
Synapse, 12
Synchronization
    estrus, 207
    parturition, 223

T

Tail-biting, 181
Taste, 20
Taxis, 279
Teasing, 207, 211
Teat-order, 82, 124, 182
Teat-seeking, 68–70, 72–3, 119, 176,
        228–9
Teleceptors, 12
Temperament, 94, 151
Tending bond, 198, 204
Tendon, severed or contracted, 260
Territorialism, 91, 124–42, 202, 279
    cattle, 169–70
    horses, 158
    pigs, 186–7
    poultry, 193
    sheep and goats, 178
Testis, 17
Testosterone, 17, 96
Thalamus, 22
Thermoregulation, 47–8, 132, 184–5
Thigmotaxis (thigmotropism), 279
Thirst, 100
    see also Water intake
Threat behaviour, 16, 91, 96, 97, 139,
        160, 169, 200
Thyroid gland, 16
Thyroxine, 16
Tinbergen, 1
Tongue-rolling, 245
Tonic immobility, 279

Toys for pigs, 181
Training principles, 39–41
Tranquillizers, 126, 179
Transit behaviour, 52–3
Transport, 52–3
    and stress, 235
Travel to grazing, 106–7, 116–17, 177
Trial-and-error learning, 39, 110
Trot, 119, 154
Trumpeting, 152
Tryptophan, 144
Turkeys, see Poultry

U

Urination, see Elimination
Urine production, 104

V

Vertebral lesions, 260
Vices, see Abnormal behaviour
Vision, 19, 31–2
    neonatal, 68
Visual area of cerebral cortex, 13
Visual stimuli, 19
Vocalizations, 82, 89, 92, 97, 136, 138,
        151, 171–2, 188

W

Walk, 118
Wallowing, 49–50, 184
Wandering, piglet, 81–2
Water availability, 106, 116
Water intake, 48, 103–5, 153, 162–5,
        174, 190
    and climate, 48–9
Weaning, 107, 246
Welfare of animals and ethology, 5–7
Whitten effect, 207
Wind-sucking, 246
Wing shoulder, 261
Wood-eating, 108
Wool-picking, 245